Practical Solar Heating Manual

With Blueprints for Air and Water Systems

DeWayne Coxon

ANN ARBOR SCIENCE
PUBLISHERS INC / THE BUTTERWORTH GROUP

Copyright © 1981 by Jordan College
Cedar Springs, Michigan

Published 1981 by Ann Arbor Science Publishers, Inc.
230 Collingwood, P.O. Box 1425, Ann Arbor, Michigan 48106

Library of Congress Catalog Card Number 81-65327
ISBN 0-250-40446-X

All Rights Reserved
Manufactured in the United States of America

Contents

Preface 4
Introduction to Solar Energy 5
Passive Solar Systems
    Air 11
    Water 13
    Solar Greenhouses 14
Active Solar Systems
    Air 16
    Liquid 18
Jordan Systems
    Air 21
        Serpentine Collector 21
        Manifold Collector 32
        Air Handling 38
        Storage 38
    Liquid 47
Appendix A: Weather Data 59
Appendix B: Bibliography 75
Appendix C: Glossary 79
Appendix D: Jordan College and Energy 85
Appendix E: Alternate Energy Installations on the
    Jordan College Campus 98
Appendix F: Solar Tax Credits 95

Preface

We live in an energy-conscious society. And that's good, because for all too long man was not as conscious about energy as he should have been.

One of the best sources of energy is the sun. We all have noticed how warm the inside of an automobile can get, even on a cold day, while it is sitting in the sun. The rays of the sun shine down on the windshield of the car and make the interior very warm, even uncomfortably so. Indeed, the sun may well be considered our most important natural resource. It is a never-failing source of abundant, non-polluting energy.

We now know how to collect the energy from the sun's rays and, just as we can store water, grain, or gasoline, so we now can "store" this energy for later use. There is, of course, a difference. Where we can store, say, water almost indefinitely, we can only store solar energy for a few days. But those few days are very important.

Leaders in the field of solar heating tell us that by 2000 about thirty percent of the nation's demand for heating and cooling needs could be supplied by solar energy. Some believe that by the turn of the century twenty percent of the **total** energy needs of the United States could be met by this same source.

With so many people asking questions about the sun and its energy, we have decided to help. Therefore, this manual. Its purpose is to give some simple suggestions about how you can obtain heat energy from the rays of the sun and store it for a few days when the sun is not shining.

In this manual we explain and then demonstrate how solar heat can go to work for you. If you can get the sun to work for you, you can save a lot of the money you are now spending on heat bills. The explanations and demonstrations in this manual are done step-by-step so that with very little prior knowledge you can solve some of the problems you may now be facing when it comes to heat.

This manual does not attempt to say everything that could be said about alternate energy sources. It makes no effort to be scientifically and technologically exact in everything it says about solar heat. It is, rather, a primer, a beginning, a start. We promise that it will be useful.

Introduction and Advice About Solar Energy

The object of all solar heating systems is to generate heat from sunlight, to distribute this heat to areas where it is required, and to store it for use in times when no sunlight is available.

The methods by which these objectives are achieved may be broadly classified under two headings—passive and active.

Any building with windows which admit sunlight has a passive system of sorts. Everything on which sunlight falls—walls, floor, furniture—becomes heated (i.e. is an "absorber") and radiates this heat into the space around it. The air becomes warm and circulates by convection, carrying heat throughout the house. When the sun sets, shades are drawn, so preventing heat from escaping by radiation through the windows. The heat absorbed by the walls, floor, furniture, etc., is now given up to the surrounding air; thus we have a simple, though not very effective, form of short-term storage.

In general it may be said that passive systems are structurally integral with the building; to add passive solar heating to an existing building is difficult and may not be very effective. However, when a building is sited to take advantage of the topography, and is designed for passive solar heating (by incorporation of, for example, a heavy masonry absorber such as a "Trombe wall" in mild climates, (or a water-filled drum wall), significant reduction in heating costs can be realized.

Active solar systems require the introduction of energy from a separate source—generally in the form of electricity. This is necessary to operate blowers and automatic dampers in an air system, or the pumps and valves associated with a liquid-medium system. Active systems can be effectively installed in an existing building, although maximum efficiency will be obtained where a new building is designed to allow for best location of collectors and storage bins or tanks.

As many people now live in homes that necessitate a retrofit solar system, in most cases that system will need to be an active system. Thus, the information in this book on what a solar system is and the various parts that comprise it. For those who would not desire to build a solar collector and storage, the information as to what a system is should help in making a wise choice of the correct system for your home.

Early in 1976 a large, wholesale heating distributor invited me, and several others, to observe a demonstration given by a prominent national solar manufacturing company. The presentation was convincingly delivered by a technician-salesman. The shiny, new, and complicated machinery was expertly designed to get the job done.

But the discussion at the close of the presentation left one important matter unresolved: the cost!

You see, in northern areas like Michigan, Ohio, or Indiana, the sun does not shine as often as it seems to shine in other states. There are so many "non-sun" days that a home heated by a solar system would need a back-up system. The back-up system would "kick in" or go on when energy from the sun was used up. So a home owner needs, in such northern states, not just a solar system, but a back-up system burning coal, oil, or natural gas.

When a company puts a $10,000 price tag on its solar system, that in itself is enough to make one gasp. But when a homeowner realizes that he needs a back-up system in addition, it is likely for him to say, "No thanks. The cost is just too great." Federal tax credits and credits in some states can now, since 1979, reduce that $10,000 nearly one-half, but to budget minded people, the cost still remains as a deterrent to purchase. Fuel oil at nearly $1.00 a gallon and rising gas costs will soon force people to consider options to traditional fossil fuels.

When the demonstration was completed, I submitted an evaluation which suggested the following: (1) It was not very likely that in the next few years people would put solar energy systems in to their homes. The relatively low number of sun-days in the north combined with the cost of a solar system plus a back-up system would simply cost too much. (2) If there is a market for solar systems, it would be in commercial installations such as gymnasiums, factories, stores and shops.

Though the sun may not shine in northern areas of the United States as much as it does elsewhere, the sun is extravagant in the amount of energy it pours out over the earth. Halary, in his book, **The Coming Age of Solar Energy**, wrote, "The population of the world consumed about 90 trillion horsepower-hours of energy in 1972. . .

while during the same 12 month period Ole Sol lavished 1.5 **million trillion** horse-power-hours of sunlight on the planet." He added that only a tiny fraction of this sun is converted by green plants into food. The rest is used in the making of fossil fuels, and that's a process which takes thousands of years.

There is plenty of sun. More than enough to satisfy everyone's needs. What man must learn to do is to convert the sun's energy directly into use by collecting the heat as it comes from the sun and storing it for the sake of non-sun hours. That way we wouldn't have to wait for tree growth for fuel or the multi-thousand year process for producing coal and petroleum. Our problem is not abundance, but storage.

Fossil fuels have, from earliest recorded history, been the primary source of energy for the human race. The history of energy needs in the United States records the use of wood, coal, natural gas and oil as our primary sources for energy. Wood has always been most abundant, and its renewal cycle of twenty years makes it our only really quick renewable fuel. Our coal supply is abundant with about a predicted 300 years' supply. Some scientists believe that, if we continue at our present rate of consumption, we will run out of natural gas in only fifteen years.

When the nonrenewable sources of energy, such as petroleum and natural gas, have been used, America will then be forced to turn to alternate forms of energy. Right now the two most obvious choices are nuclear power and solar power. But with environmentalists making us all more cautious about the potentially bad side-effects of nuclear power, its use on a broad scale may be limited. The sun, then, becomes our greatest asset. Its energy potential clearly elevates it above other alternate energy forms such as water, wind, and biofuels.

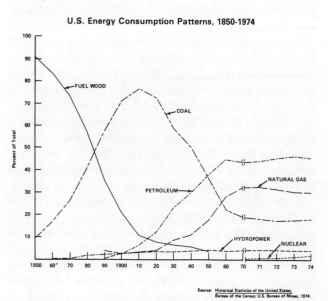

Efficiency of Energy Utilization, 1960-85

Source: U.S. Department of the Interior, 1974.

U.S. Energy Consumption Patterns, 1850-1974

Source: Historical Statistics of the United States, Bureau of the Census; U.S. Bureau of Mines, 1974.

It is interesting to note that history records many instances of solar use. Legend has it that Archimedes destroyed the Roman fleet at Syracuse, in 212 B.C., by using a magnifying glass to concentrate the sun's rays, thereby igniting the wooden ships. Magnifying glasses have often been used to set fires. Scientific solar research was conducted by Galileo in the seventeenth century. In the eighteenth and nineteenth centuries scientists used the sun to heat engines and irrigation pumps. In the eighteenth century de Saussure invented a solar cooker by using a wooden box with the inside painted black. The information he provided, enabled scientists worldwide in their testing of darkened tilted boxes covered with glass to capture the sun's rays. One can read of experiments in India, Israel, England, France, Germany, Egypt, and the United States being conducted to discover new ways to utilize the sun's energy.

America has made herself responsible for an unprecedented number of energy problems. Back in 1952, for example, a report was submitted to President Truman which projected, by 1975, a market for 13 million solar-heated homes in the United States. Unfortunately nobody listened. President Carter, in his energy plan of April 18, 1977, projected the hope that 2½ million homes would be solar heated in the next decade. In contrast, two scientists in 1961 concurred that there were fewer than 20 homes in the **world** heated by solar energy.

Why wasn't anybody doing anything?

Part of the reason lies in our unbounded optimism regarding the price and supply of fossil fuel. In 1964 Daniels wrote, "Combustion fuel in the United States and elsewhere is so cheap that it is difficult, except in special cases, to save enough fuel by using the sun to compensate the large capital investment required for the solar heating system."

How times have changed. Bethlehem Steel, a giant in the use of fuels, expressed its corporate concern in February, 1977, when it said, "The mideast oil embargo in 1973-1974 meant higher prices, gas lines, more unemployment, and more inflation. At that time we imported 38 percent of the oil we consume. Today America imports 42 percent and the OPEC nations just raised prices of oil again." Bethlehem actually paid for two full pages in **Sports Illustrated**, one of them left blank for people to write on, in support of President Carter's energy policies. And President Carter, in his inaugural speech (1977) made energy his number one priority!

All these concerns have still not resulted in the majority of Americans realizing the problem. A recent Gallup Poll showed that less than half the American people think there is a real danger of running out of oil or natural gas in the 1980's, as predicted by experts. On April 19, 1977, the White House released a CIA report indicating that the world's oil reserves probably will be insufficient to meet the energy demands of some countries by the mid-1980's.

Adding to these Washington priorities are the forecasts of meteorologists that weather trends will continue colder until the year 2000. "We can expect a cold phase to begin by the early 1980's that should result in some of the coldest temperatures in this century", Harold Bernard said, a meteorologist from Environmental Research and Technology, Incorporated, in Concord, Massachusetts.

The whole world has now become aware of the impending energy emergency that shut schools and closed factories in 1976-1977. The most optimistic energy experts have predicted that the energy problem will only get worse before it can get better. Government analysts predict that with natural gas prices deregulated the cost will triple. The Department of Energy has stated that soon the cost of a BTU will be the same whether one uses gas, oil, or electricity to heat. The unstable condition of the Near East and the political conditions with Iran only augment our precarious fuel supplies.

According to the Stanford University Research Institute, Americans during the 1960's used increasing amounts of energy to heat their homes. In fact, total fuel energy consumption rose from 4,848 trillions of BTU's to 6,675 trillions of BTU's. Unfortunately much of the trillions used went up into the atmosphere, wasted due to poor insulation or chimney draft.

Many home owners in the sixties tried to beat the heat problem. They were encouraged to build their homes with most windows facing south. They later realized that the heat loss during sunless days and long winter evenings negated the benefits of whatever sun they did get.

Most people who are presently suffering from rising utility heat bills do not have the luxury of placing their homes at ideal elevations to secure minimum heat loss. They cannot bury the northwest wall in a constant 55°F earth-bank and allow the front of the house to face south for optimum solar reception.

Data from the U.S. Weather Bureau for cities in the tri-state area of Ohio, Indiana, and Michigan list all major cities as lying in the 50 percent to 60 percent area for total sun days. This means that sun days do not exceed about 180 to 190 per year. Disappointed? Remember, the sun can be collected on many hazy days not officially classed as sun days.

The cities in Michigan, Indiana, Illinois, Ohio, and Pennsylvania also need approximately 6,000 degree days heating, as listed in the **ASHRAE Guide and Data Book.** Degree days heating simply means adding the number of degrees that the temperature falls below 65°F. This is done daily and totaled at the end of the heat season. For example: Anchorage, Alaska needed 10,864 total heat degrees to keep buildings above 65°F. On the other hand, Honolulu, Hawaii needed zero (0) degrees because the temperature never dropped below 65°F.

Any person considering solar heat should study the climatology data for the area where his collector will be placed. The abbreviated information in the preceding paragraph gives some insight into the kind of information data. Look in your local library for the **ASHRAE Handbook of Fundamentals.** The local office of the National Weather Service can also give high and low temperature listings to interested individuals, as does Appendix I of this book.

The tilt of the earth in relation to the sun makes it necessary for us to locate the solar collector in such a position that we gain the most direct line to the sun's rays. Since Michigan, Indiana, Illinois, Ohio, and Pennsylvania are approximately at the forty-fourth latitude, the correct angle for the collector is achieved by adding 15 degrees of tilt. Thus, 44 + 15 = 59 degrees. People wishing greater fall and spring generation are building collector slopes at 45°. This tilt on the collector will vary a few degrees from state to state. Exact latitude of your area can be secured by checking the ASHRAE tables in the back of the book.

The decision as to the size of your collector should be made in relation to the size of your building, weather conditions, and insulation. Manufacturers of flat plate collectors vary in the amount of square footage needed. Absorber material accounts for some of this disagreement.

Some manufacturers recommend that for 100 percent efficiency, 50 percent of the square floor footage of the building must be the size for the collector. But that would make for one huge collector in relation to the size of any building. Fortunately most manufacturers believe that 25 percent of the square floor footage is the most efficient, especially when figuring the cost per square foot of collector versus dollars saved in heat. Since few manufacturers will guarantee 100 percent solar heat efficiency, it is important to determine how much collector would be economical in proportion to the total reserve heat than can be stored.

What about cooling? Tests that have been conducted in the mid-west area show quite convincingly that present solar systems are not very practical for cooling purposes. Summer evenings in the mid-west simply do not usually drop in temperatures enough to cool the storage area so that the cool air could be distributed the following day.

If one were living at higher elevations, things would be different. Substantial degree differences during evening hours can sufficiently cool a storage area if the cool night air is drawn through. Some experts have even run an air conditioner into the storage area at night, and have drawn off the cooler temperatures from storage the following day, saving electricity in the process.

The sun is the only limitless energy supply this world has. Its non-pollutant rays qualify it for harnessing in the twentieth century.

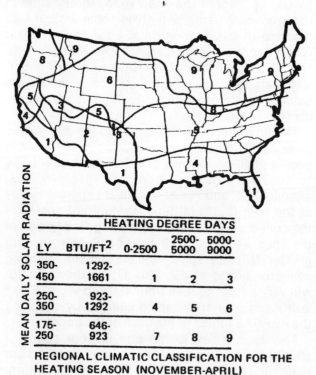

REGIONAL CLIMATIC CLASSIFICATION FOR THE HEATING SEASON (NOVEMBER-APRIL)

You should also realize that the orbit of the earth around the sun is not in a circle, as many people think, but is actually an eclipse. The earth approaches the sun at a distance of 89.8 million miles in December, when the earth is closest to the sun. The earth is farthest from the sun, some 95.9 million miles, in June. So why for us is it hot in June and cold in December? That has to do with the axis tilt of the earth and other orbital factors. And that, too, explains why December is warm in South America, and June is cool.

Now that we understand why we have such good hot sun in December, we need to know how to maximize on keeping it. So, you need to know a little about insulation.

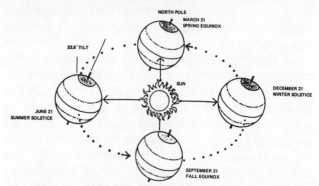

To get right to where we live, we realize that buildings are artificially created environments. Inside the temperature may be significantly different than outside.

What you may not have known is that builders did not realize the value of proper insulation until the 1950's. It was the advent of electric heat in the home-building industry which caused an awareness of how important insulation is.

Until then, many homes were not properly sealed off from the environment to maximize on conserving heat. When wood and coal were cheap, we pushed the cold out by adding fuel to the fire, or by adding a stove to the living room in addition to the kitchen. Even with the widespread use of heating oil or natural gas, the prices for these utilities were usually low enough so that people just didn't think much about insulation.

Ah, how times have changed! Coal has quadrupled in price, petroleum has doubled, and gas is expected to rise at the rate of 15 percent each year for the foreseeable future. No wonder we are being told repeatedly that **the cheapest way to heat is to insulate!**

New insulation methods, such as liquid polyurethane foam, have entered the market to help seal the inclement elements out and keep the heat in. Even utility companies are insulating homes, and allowing people to conveniently repay the cost through adjusted monthly fuel billings. The U.S. government is considering making ceiling insulation mandatory in order to cut the total consumption of fuel. Fuel economists are convinced that part of the energy solution is to stop the heat already in our homes from getting out into the atmosphere.

We Americans are more insulation-conscious than at any previous time in our history. Any person now considering solar heat must first realize the great savings that will inevitably come through energy conservation. We must give proper attention to doors, windows, and chimney drafts that steal heat from our buildings.

New home owners and builders are now acutely conscious of air currents. They want to take advantage of these for cooling homes in summer and protecting buildings in winter. Shrubs and trees are now being strategically placed about buildings and homes to control and better use the wind and the sun. Some homes are being built into hillsides to protect the north and west exposures, and make maximum use of an earth which gives off a steady 55°F.

Thermopane windows have become, in spite of their initial expense, absolutely essential to energy minded builders. Storm windows are being added to homes without welded glass. Sun screens are being used to reflect heat inward and draw cold outward, and are being reversed in the summer for cooling purposes. Infra-red sensors are now used to detect heat loss before and after a building has been insulated. Maximum insulation is being used in sidewalls and ceilings to deter heat loss.

All of these factors must be considered, and many of them implemented, before people can truly maximize the benefits of solar heat. It makes little sense to put in a solar system, even if done relatively cheaply, if the solar energy stored for non-sun hour use would be lost due to improper insulation!

Insulate The House

Heating Zone	Recommended for		
	Ceiling	Wall	Floor
0,1	R-26	R-13	R-11
2	R-26	R-19	R-13
3	R-30	R-19	R-19
4	R-33	R-19	R-22
5	R-38	R-19	R-22

Insulate The Solar Energy System

Liquid Systems
Piping   R-4
Storage (Water) R-11

Air Systems
Ducting   R-6
Storage (Rock) R-19 (Indoors)
                R-30 (Outdoors)

Passive Solar Systems:

Air

Industries and homeowners alike are searching for economical ways to beat rising energy costs. Whereas, in the years preceding the 70's, building architecture made little or no use of energy conserving features such as insulation and window placement, there is now a growing understanding that building design and construction must take advantage of alternate energy sources, foremost among which is solar energy.

As was briefly mentioned in the introduction to this manual, solar energy systems in general fall into one or other of two main categories—active or passive. An active system uses an additional energy source (generally electricity) to operate the blowers, pumps, valves and/or dampers necessary to circulate the heating medium. Passive systems, on the other hand, use the entire building as a collector and storage unit. They rely on the design of the building and the materials used in its construction to capture and store the sun's energy—or to shut it out when it is not needed. During the winter, heat is distributed by non-mechanical means, i.e. by conduction, convection and radiation. During summer, the design of the building and an assortment of shading devices prevent sunlight from entering and thus keep the interior cool and comfortable. As with active systems, a passive system usually includes a back-up heating system for use during extended periods of overcast or cold winter weather.

There have been many technological advances in energy conservation—many so impressive that we might be tempted to congratulate ourselves as modern geniuses. But to be honest we should strip away such pride and admit that many of our "modern" techniques are really not so modern after all.

Passive solar energy is not a new discovery. American Indians built entire adobe brick villages a thousand years ago using passive design principles. Lacking modern technology and out of necessity, they made use of building characteristics which have become the core of passive design: architecture related to site, climate, local building materials and the sun.

Architectural designs have changed considerably since adobe cliff dwellings, but ironically, architects are admitting that ancient building designs offer many intelligent alternatives to conventional construction. Engineers, scientists and industry are now trying to duplicate these old ideas—but with 20th century adaptations. They are succeeding in their goal, but not without more than a few obstacles, some of which are technical, some psychological.

Although the prospects for full national exploitation may not be exactly sunny, the future is far from cloudy. Indeed, there appears to be ground for enthusiasm. Interest in passive is growing, training programs and factual energy data information are being developed, and research institutes are beginning to give solar development the priority it merits.

An engraving of Pueblo Bonito.

As buildings appreciate in value, passive systems will, too. Costs will be justified as energy conservation and as an investment in the future. Passive design should pose no problem to the construction industry because most plans call for standard building materials. As energy costs rise, the economics of passive solar energy systems will become more favorable; the prospects for passive solar energy will improve.

Certain traditional problems must be overcome when designing passive solar systems: heat loss through glass, glare, placement of adequate thermal mass, and abnormally large temperature swings between rooms. Although they have long plagued engineers, these problems do not present the difficulty that passive designers once faced. Products are emerging as solutions which make solar energy a reasonable alternative to conventional energy sources.

Passive solar energy systems are composed of four major elements: solar collection, heat storage, distribution and control, and retention.

South-facing windows are the major ingredient for solar collection. Transparent glazings, such as glass or synthetic glass substitutes, can be designed to attract the sun's heat, prevent it from reradiating, and reduce glare on the inside. Triple pane windows are becoming common on the market.

Once heat has been collected, it must be stored until it is needed. To accomplish this, floors, walls and ceilings double as both structural supports and thermal mass for heat storage. The essence of the mass (masonry or water-wall), thickness of the sides, and use of thermocirculation vents to force air past the thermal mass, are variables that can be regulated to control indoor temperature fluctuations. Insulating shutters or other movable insulation can further reduce temperature fluctuations to about 10 degrees. Shading devices play the important twin role of retention and reflection of heat in passive solar energy systems. Beadwall systems, insulating shutters/curtains/shades, exterior solar shades and screens, awnings and other products have been established as an economical way to control solar heat gain and loss.

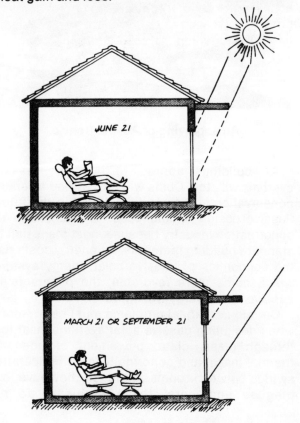

SOUTH WALL WINDOW — Overhang Shading

If the shading device can double as a reflector or movable insulation, so much the better. The effectiveness of a shading device depends on how well it shades the glass in the summer without shading it in the winter. A goal of passive solar design is for the large south-facing windows to allow maximum solar gain in the winter. Without shading devices or movable insulation, the same result would occur in the summer and cause overheating. Thus, movable insulation serves a dual function. First, the insulation covers the solar collector to reduce heat loss on winter nights. Second, it serves as a barrier to heat gain in the summer, thus keeping the building cool.

Three basic design concepts form the core of most passive solar heating systems: direct gain, indirect gain and isolated gain. These concepts involve different methods of collecting and distributing the sun's energy so it functions as a heat source for a living space.

The original and simplest approach to passive solar heating is the concept of direct gain. With this system the actual living space is directly heated by sunlight filtered through an expanse of south-facing glass. The floor, walls or ceiling of the living space serve as strategically located thermal mass for heat absorption and storage. The building becomes both a live-in solar collector and heat storage and distribution system.

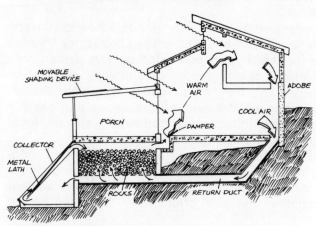

PASSIVE HEATING SYSTEM DIAGRAM — Air Flow

With indirect gain, the thermal mass is located between the sun and the living space. A trombe (thermal) wall absorbs the sunlight, stores the heat, then converts the heat to thermal energy and transfers it into the surrounding area.

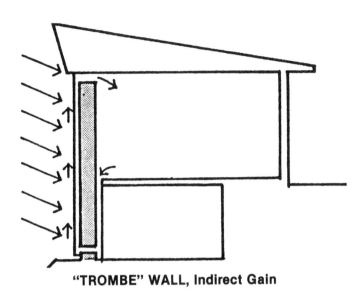

"TROMBE" WALL, Indirect Gain

The third approach is called isolated gain. Here solar radiation is collected and stored in a separate partition from the living space. Convection vents located near the floor and ceiling of greenhouses or sun porches allow the heat to be transferred from the solar heating area to the living space as needed.

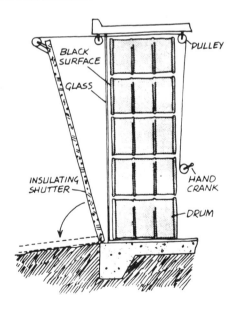

INSULATING SHUTTER

While obstacles may delay progress temporarily, the advantages of passive solar energy cannot be ignored. Passive solar energy places a low burden on the earth's resources. Because the design is simple there is little chance of operational failure. Little technical maintenance is needed, and what is required, the owner can usually handle himself.

Passive systems seldom require equipment certification. Most construction materials are readily available and can easily be installed. Because they do not operate mechanically, passive systems are noiseless. They are frequently invisible from the interior because they have no radiators or fans. They are low-cost in the long run.

Depending on climate, passive design can provide much of a building's energy needs. Passive solar energy systems can be incorporated into new construction and many times retrofitted to accommodate present buildings. Its impact is just being felt. Its potential and scope are just beginning to be realized.

Water

Passive solar hot water heaters were used in the southern parts of the United States prior to the natural gas era. One can see the rusty remains of them in cities such as Orlando, Florida. The modern nation that has caused a recent sensation in passive water heaters for domestic use is Israel. Over 80% of its residents in cities use the simple collector-storage tank design for domestic hot water. Collectors on roofs are as plentiful as television antennas. The systems simply circulate water from the collector to storage and then take the hot water from the tank to the house. Open systems in Israel also use a float-valve while closed systems, if using anti-freeze in the United States, use a heat exchanger.

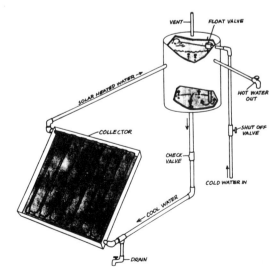

PASSIVE WATER SYSTEM

The passive water wall collector utilizes a blackened absorber midway in a closed horizontal storage. The tank on top of the collector supplies water for exchange into the collector. Connection to the domestic hot water tank or living space can be by heat exchanger or movable insulation.

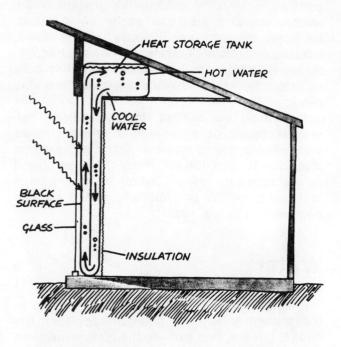

Southwestern people have been using the integrated collector and storage with water filled cylinders with some success for several years. These are of value where not only domestic hot water is needed but cooler nights demand a lighter heat load. As water will store more BTUs per pound, the cylinders when blackened, store and then transfer the heat passively to the house.

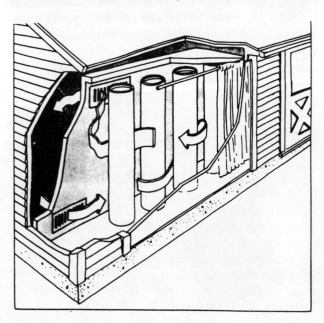

INTEGRATED COLLECTION AND STORAGE –
           Water Filled Cylinders

Solar Greenhouses

Traditionally, northern climates impose short growing seasons, and unseasonal frosts can bring additional problems.

Greenhouses lengthen the growing season, but in return demand heat to sustain the plants during periods of no sunshine. Commercial growers budget a substantial cost factor for fuel to heat their greenhouses. Single pane glass and little or no insulation also contribute to the problem.

Experimentation in greenhouses adapted to northern climates has produced encouraging evidence of low-cost heating for users of such structures. The temperature in the solar greenhouse at Jordan College in Michigan never fell below 32ºF during the years of 1978 and 1979. No auxiliary heat was used in the greenhouse, and on sunny days, while the ambient temperature was well below zero, the temperature in the greenhouse rose to nearly 100ºF.

Greenhouses for northern climates must be designed to conserve all solar and ground heat, particularly through the long sunless days and cold nights of winter. This process is accomplished by applying several simple rules. If these are followed, the growing season for vegetables and flowers can be lengthened considerably, and fuel bills can be drastically reduced.

Attention must be directed to three areas when constructing the greenhouse. Footings, and all walls below grade must be insulated; glazing must be double or triple-layered, and thermal mass must be created to store the heat from the sun's rays in accordance with the best passive design principles.

Active Solar Systems

Active solar heating installations usually comprise four component systems; the collector, the heat storage system, the circulation and distribution system, and the control system.

Collectors may generally be classified as flat-plate, concentrating, and semi-concentrating types. Concentrating and semi-concentrating collectors produce much higher temperatures than are usually required for domestic purposes, and they are expensive. The solar heating systems described in this manual, both air and liquid, utilize flat-plate collectors. The efficiency of this type is well-established, their construction is not difficult, and they are relatively inexpensive.

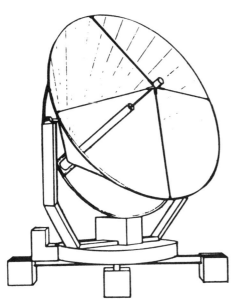

CIRCULAR FOCUSING COLLECTOR

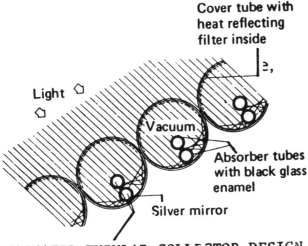

EVACUATED TUBULAR COLLECTOR DESIGN

The collector is essentially a shallow box containing a heat absorber. The box should be fully insulated, and carefully sealed so that no air can leak in or out. This is important, since such leakage would cause heat losses which might considerably reduce the efficiency of the system.

Absorbers used in flat-plate collectors are customarily flat or corrugated metal, painted (or anodized) black. If a liquid system is being used, the liquid passes through tubes attached to the metal. Sandwich absorbers are manufactured that have integrated liquid tubes in the metal.

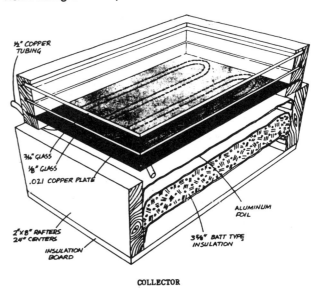

COLLECTOR

The collector is covered with a substance which will transmit the maximum amount of sunlight possible, while only permitting the least possible amount of heat to escape by radiation. Various plastics, fiberglass or glass are used, glass being the preferred substance. Glass transmits a greater proportion of incident light than plastics, is more durable, and permits less loss by radiation; however, its high initial cost and susceptibility to breakage may cause it to be passed over in favor of the less expensive alternatives.

The Storage System consists of a well-insulated container and a suitable heat-absorbing substance, such as rock or water. The volume of storage material needed, whether solid or liquid, is determined by the total heat requirements of the building in which the system is installed. This heat requirement, which is related to the floor area of the building, also governs the size of collector necessary.

There is thus a relationship between building floor area, storage volume and collector size; for most installations located between 40 degree and

50 degree North Latitude, this relationship may be expressed as follows:

Collector Area = floor area of building / 4

and

Storage Volume = 2 gallons liquid or 0.30 cu/ft of solid material per square foot of collector area.

Thus, for a well-insulated building having a floor area of 1200 sq. ft., a collector of 300 sq. ft. surface will provide sufficient heat energy to serve a 600 gallon storage tank, or a rock bin containing 90 cubic feet of stones. These figures are, of course, approximations, but will serve as a useful guide in the design of a solar heating system.

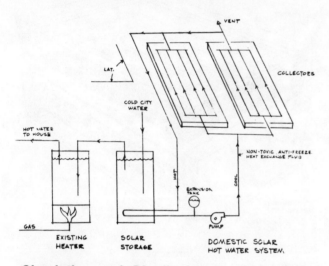

Circulation and Distribution — If the solar heating system is being installed in an existing building, the circulation and distribution equipment already in place may be used. The storage bin (for an air system) or tank (for a hydronic system) should, in such a case, be located close to the existing central heating unit. The solar heated air, or water, can then be arranged to feed into the air ducting or plumbing installation.

Where the system is being installed in a new building, ducting and/or plumbing should be designed to accommodate whatever auxiliary sources of heat are intended for use.

A National Review of Air Systems, Liquid Systems: Both Work Well, Both Require Some Special Considerations

Air and liquid systems each have their own special advantages. It is up to the builder to choose one or the other depending upon his needs. Both types of systems will work well when properly designed and installed. Quality workmanship is essential for good performance regardless of the type of system used.

In most cases, adherence to the recommendations contained within HUD's **Intermediate Minimum Property Standards: Solar Supplement** *will help avoid many installation problems. However, there are some areas of special concern for each type of system. The builder should be aware when special attention is needed for air or for liquid systems. And when combined with some of the other technical resources, builders, installers and manufacturers have a comprehensive package of information to assist them in securing trouble-free operation in whatever type of system they choose.*

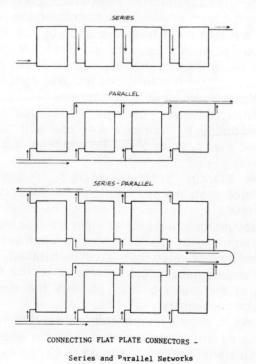

CONNECTING FLAT PLATE CONNECTORS —
Series and Parallel Networks

Air System Experiences

Air systems are increasing in popularity for several reasons. They are simple to build on-site for those without plumbing skills. They should be less likely to encounter water damage, freezing or corrosion problems. And they can be connected fairly easily to existing forced air heating systems.

Some specific areas of concern for air systems include:

1. Air Leaks—Air leaks can drastically reduce system performance and create excessive noise. Leaks can be external or internal. An external leak releases solar heated air to the outside or to conditioned or non-conditioned space. Internal leaks release hot or cold air from one part of the system to another in the wrong way or at the wrong times.

External Leaks—External leaks occur through joints and penetrations in collectors, storage bins and ducts.

• Leaks can be prevented by sealing all joints

with a permanent, flexible, high-temperature sealant.
- Wood should not be exposed to excessive heat and air flow as it will dry and crack.
- Collectors and ducts should be supported to prevent vibration and sagging which could open joints.
- Flexible connections should be used to prevent excessive vibration.
- Round ducts with slip fit joints are easier to make leak tight than rectangular ducts connected with slips and drives.

Internal Leaks—These leaks occur through installed dampers. Damper leaks have occurred because:
- back-draft dampers were installed backwards;
- blades were loose;
- blades or linkage were binding or damaged;
- non-gasketted dampers were used;
- operators were out of adjustment;
- dampers were too small for the duct;
- dampers were poorly fitted into the duct.

Damper leaks also reduce performance by allowing the mixing of hot and cold air, by passing elements of the system and allowing cold air from non-operating collectors to flow through domestic hot water coils and freeze them, causing water damage. Automatic dampers should be the low-leak gasketted type. Manual dampers should be similar or should be the slide type with tracks for ensuring a tight fit.

2. **Air Balance**—
- A reverse return system should be used at the collectors to aid in balancing.
- Balance dampers should be installed at tee fittings and may be required at individual collector taps depending on duct size and layout. Balance dampers should be fixed and marked after balancing.
- The storage bin design should allow for stratification of heat, and distribution of air over all the storage media.
- A means of bypassing storage for summer domestic water heating and for winter auxiliary heating should be used.

3. **Flow Rate**—Excess pressure drop is the greatest cause of low flow rates and also results in noisy systems. Fan size, duct size and length, filters, coils, dampers, collectors and storage bin should be carefully evaluated during design, and installation should not vary from the design significantly.
- Gradual duct transitions and turning vanes should be considered.
- Filters and coils require fairly large transitions to reduce static pressure losses at these locations.
- Gravity dampers installed backwards and loose duct lining can cause excessive pressure drop as can dirty filters and coils and stuck dampers.
- Inadequate space for equipment and ducts commonly results in poor fittings and duct constrictions.

4. **Auxiliary Equipment**—
- Some manufacturers of heat pumps recommend maximum air temperature flowing across the inside heat exchanger coil while the compressor is operating. It may be necessary to lockout compressor operation when heating directly from the collector.
- There have been some instances of excess heat loss occurring when a fossil-fueled furnace fan is used to distribute heat from the solar system while the burner is off. This is caused by a stack effect through the furnace heat exchanger up the flue removing heat from the system. Automatic flue dampers, where approved, might be considered to reduce this problem.

5. **Access, Identification, and Operating Instructions**—In some instances, filters, dampers, controls and operators are inaccessible and/or identified for the solar home occupant. And, in some cases, no instructions have been furnished. Quite frequently, there are manual by-pass dampers which must be seasonally adjusted by the occupant, and these must be identified.

6. **Health**—Although no health-related problems have been experienced to date, precautions should be taken to eliminate air contamination and flame/smoke spread.
- Loose fiberglass insulation should not be exposed to air flow.
- Wood and foam insulations in the air stream should be protected with fire-resistant coverings to prevent the spread of smoke, flame, and toxic fumes in the event of a fire.

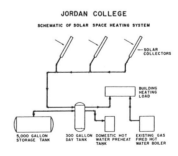

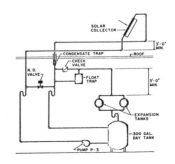

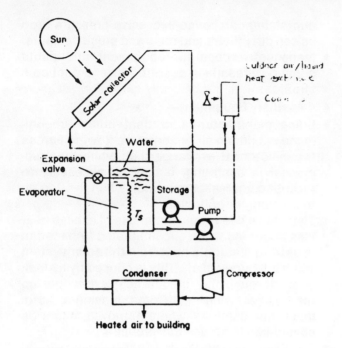

SOLAR COUPLED HEAT PUMP

(Water to Air Heating Mode)

Liquid System Experiences

Liquid systems have been the choice of many builders for a variety of reasons. Liquids have higher specific heat properties than air, therefore requiring less storage volume and pipe diameter. They can also be connected easily to existing forced air heating systems. And in some cases, they can be used without heat exchangers, such as when domestic water is used as the transfer fluid.

Builders should be aware of following concerns when installing liquid systems:

1. Water Leaks—Water leaks can drastically reduce system performance as well as damage property. Leaks most commonly occur in joints in the piping of liquid systems, but can also occur in storage and in other subsystem components.

Joint Leaks—
- Soldered joints must make allowance for some thermal expansion/contraction and vibration. High temperature solder, such as 95/5 tin/antimony, should be used.
- Threaded joints should be made up tightly, using pipe dope or tape. The joining of pipes of dissimilar metals, such as copper and aluminum, should be avoided.
- Mechanical joints, such as compression or flare fittings, should be made up carefully and tightly. Hose clamps should be non-serrated and installed carefully to avoid puncturing flexible connections.

Storage Leaks—
- Plastic and concrete tanks require much care in design and installation to avoid problems caused by settlement, such as cracking. Concrete tanks should be lined with a suitable heat-resistant, non-porous material such as fiberglass or butyl rubber.
- Plastic and fiberglass tanks must be designed to withstand the anticipated temperatures and pressure.
- Steel tanks used in open systems should be lined to avoid oxidation. The use of dissimilar metals in tanks or their connections should also be avoided.

Evaporation—
- An open loop system can lose water through evaporation. In these systems, an automatic fill or low water indicator should be used.

2. Freezing—Freezing can cause severe damage to the solar energy system and the home. It can result from poor installation procedures, equipment failure, or operational errors.
- Storage capacities in draindown systems must be sufficient to hold *all* system fluid to avoid exposing any fluid residue to freezing conditions.
- Pipes, including the collector manifold, must be pitched to allow proper drainage. Vents must also be included in proper locations to aid in draining.
- In draindown systems, freezing can result from malfunction of the solenoid valve, temperature controller, or sensor. Incorrect settings of sensors can also allow freezing to occur.
- The antifreeze solution can become diluted by the automatic make-up of system liquid with city water.
- Inadequate, damaged, compressed, or wet insulation will lose its resistance to cold infiltration. Freezing may result. Insulation should be protected from sun and water if exposed, and from compaction and water if buried.

3. Low Flow—Low flow rates can be the result of obstructed pipes, poorly sized pumps, or equipment failure. Because low flow rates lower collector efficiency, and therefore system performance, care should be taken to ensure proper flow rates.

Pipe Blockage—
- Pipes or collectors can become blocked by foreign objects. All pipes should be clean before assembly. Keep all tanks, pipe ends, and fittings closed with clean caps until assembly. Piping should be flushed after installation is complete. Strainers should be installed upstream of pumps.
- Blockages can be caused by scale or corrosion.

- Air blocks can form in pipes. Proper pipe pitch and placement of air vents can help avoid this problem.
- Control sensors or their wells should not protrude too far into piping.

Wrong Pump—
- Pump capacity depends on head. The use of an incorrectly sized pump can result in low flow.

Valve Malfunction—
- Valves may become stuck in shut position, or, in the case of three-way valves, in by-pass position. In general, configurations using two two-way valves have given better results than those using one three-way valve.

4. Corrosion—Corrosion can reduce system performance and cause damage to the system or the home. Careful material selection, attention to installation details, and proper maintenance routines are called for.
- Avoid contact of dissimilar metals.
- Use a working fluid that is compatible with solar system temperatures, pressures, and materials (joints, valves, sealants, expansion tank diaphragm, etc).
- In open loop systems, use pumps with bronze or stainless steel impellers and bodies.
- Some liquid transfer media can degrade shingles. Drainage from pressure relief valves should be planned to avoid this problem.
- Water in open loop systems should be checked for hardness and pH to avoid scaling and corrosion.
- Test antifreeze solutions if stagnation has occurred, since acids can form at high temperatures.

5. Other Operating Concerns—
- Pressure relief valves have popped off of systems, leaking working fluid and causing down time. Causes have included faulty equipment, incorrect valve relief setting, inadequate expansion tank size or location, and pipe blockage.
- System operation during periods of no sun has occurred due to incorrect control settings, poor controls, and thermosiphoning. Check valves will correct thermosiphoning.
- Noisy system operation has resulted from air being trapped in the system, vibration, thermal expansion/contraction, and excessive pressure drops or flow rates.

6. Access, Operation, and Maintenance—All components should be accessible for repair or replacement. Operation and maintenance instructions, and schematics of piping and controls, should also be in accessible locations.
- Piping should have drains at low points and air vents at high points.
- Any component which may require service should be isolated with shut-off valves to avoid having to drain the entire system during service.
- Methods for visual inspection, such as site glasses or thermometers for control accuracy, should be included in accessible locations.
- Valves should be tagged for easier operation and maintenance by service personnel or the occupant.
- Strainers should be cleaned periodically, particularly during the initial start-up phase.

7. Health and Safety—Although no health-related problems have been experienced to date, precautions should be taken to avoid scalding and potable water contamination.

Scalding—
- A tempering valve should be used to keep domestic water supply from exceeding design delivery temperature.
- A pressure relief valve should be piped to drain.

Potable Water Contamination—
- Either double-walled heat exchangers or non-toxic transfer media must be used. Systems with non-toxic transfer media should be tagged to avoid the possibility of being filled with toxic substances.
- Backflow preventers should be used to avoid mixing of transfer medium with potable water supply.

Jordan Solar Systems:

Air

The first three years of energy education programs from Jordan College resulted in over 10,000 people taking classes or receiving information. Over 3,000 people have received the building of a Jordan solar system in mini-courses or semester classes at the Jordan Energy Institute.

Two types of collector boxes for air systems are presented in this manual. The **Serpentine** pattern that draws the air from one end of the collector through a baffle series to the other end is best used where it is not possible to draw the warm air from the top of the collector. The **Manifold** design, taking the air from the bottom to the top can be used in both a passive or active mode. The particular application should determine the choice of collector box.

    A. Serpentine Collector Style
        (69 degree pitch)
    B. Manifold Collector Style
        (50 degree pitch)

Those who have the benefit of observing demonstrations of system building in class will find that setting up their own system is greatly simplified. If the construction designs described are followed, step by step, a solar system will result which meets all performance expectations. Similar systems have already been built in great enough numbers to guarantee success.

In setting up the collector, remember that it must face south. A conversion table (below) demonstrates the degree to inch conversion.

Conversion of Pitch to Degrees

Angle in Degrees	Pitch in Inches per Foot
35°	8.4
40°	10.1
45°	12.0
50°	14.2
55°	16.7
60°	19.6

The mid-western United States lies between the latitudes of 40 and 50 degrees. The number 15 can be added to the latitude; thus Grand Rapids, Michigan, at a latitude of 44° adds 15, so the total is 59°. One degree, more or less, is not critical to the success of the collector. Several collectors have been built at 45° angles to capitalize on every fall and spring solar gain.

Find a large enough space for the building of the collector; the space should be at least as large as a one-stall garage. The units should be pre-assembled on the floor before being carried to the location where they can be assembled.

It is now time to coordinate your reading with the construction of your solar heating system. Collector construction step one should be read slowly and carefully. Following the choice of your collector comes the storage diagrams. Finally, air handling modes provide the information necessary for you to activate your system.

Serpentine Style Collector

Step I

The actual construction of the collector should be preceded by reading the parts list under the construction details at the lower right hand corner of the page. That listing gives the number of 2 x 4's (24) and plywood (6 sheets) plus nails needed to build the 4 x 8 sections for the collector backing. The plywood used for sheathing on a new building also serves as a foundation materials for the collector. Step one emphasizes the necessity of keeping the outside edge of the 2 x 4's flush with the edge of the plywood on all four sides.

Step II

The parts list on assembly two shows the addition of the 1 x 12 boards used as framing for the collector. After the six 4 x 8 units are nailed together the 1 x 12 ties the units together and forms the solar box that houses the collector. Collectors that are roof-mounted should be transported in 4 x 8 sections and nailed together and forms the solar box that houses the collector. Collectors that are roof-mounted should be transported in 4 x 8 sections and nailed together on the roof. Two important construction details should be noted. The structure should be square as determined by placing a square in the corner, and the bottom edges of the 1 x 12 framing boards must be flush with the bottom edges of the 2 x 4's.

Step III

Cutting the three openings through the plywood is not difficult. If you do not have a compass to scribe the round holes, simply measure out four inches from the nine inch cross. You can throw away the round plywood but save the 4" x 18" cutout as a louver door. Care must be taken in stapling the aluminum foil to the plywood and sides. Air from the blower will tug at the corners and frayed edges, and will raise the aluminum. Staple well. Note carefully the location of the 2 x 2 absorber framing. The top 2 x 2 baffle is shorter than the bottom; the reason for the difference is to force the air through the collector evenly.

Step IV

The five choices as listed in the lower left of the blueprints all presuppose the use of flat aluminum at both ends of the collector. The use of this material at the two ends was made realizing that extrusions and other aluminum or steel fabrications would be hard to cut and nail down over an area that did not have a center nailing surface. The choice of how much extrusion to use is left to you. You will collect heat in proportion to the amount of extrusion used, up to 2½ times that transmitted by flat plate alone. If you place the extrusion at six or twelve inch intervals you must then lay full aluminum sheeting under the extrusions.

Step V

The baffles and cover supports are necessary to direct air flow and support the cover. They must be nailed in order that the cover can rest firmly on their strength. Air baffles 2 and 4 must be built so that the air directors are indented four inches from the cover band. This is very important as it will allow hot air during the summer to escape through the external hot air exit door. The 4" x 18" external air exit door should be finished with a one-inch trim to avoid air loss. The motorized arm should be adjusted to close firmly but not to exert pressure on the control mechanism.

The baffle and cover supports are nailed on two feet centers with the top of the wood flush with the sides. This will allow the covers to rest flush with the total unit. Five one-inch holes are to be drilled near the baffle cover supports. They should be plugged from the outside with low temperature foam or a waxed cork that is gently pressed into the hole. In the event that there is a power failure, the heat in the collector will force the cork or styrofoam out and allow the absorbers to cool by gravity.

Step VI & VII

The collector cover is built in six sections as the collector base was. The fiberglass cover should be drawn as tightly as possible by stapling the bottom and then pulling it tightly at the top. Silicone sealant must be applied before cover stock is stapled. When all covers are completed, lay them on the collector and use 1½-inch flush finish wood screws. Screwing them down will allow for quick replacement if damage is incurred in coming years.

Step VIII

Nailing the damper and adaptor housing to the collector completes the construction. Sealant needs to be applied to the **inside** of the wood housing before the louver is installed. The louvers must be at 90°. The shutters must not be allowed to open, even slightly, as night air will cool the store bin. Be sure your collector does not settle after positioning the louvers. Some people let the louver lean backwards at about a 2° tilt to insure complete closure. You may wish to remove the metal adaptor housing and check the dampers occasionally. Mount the adaptor with screws.

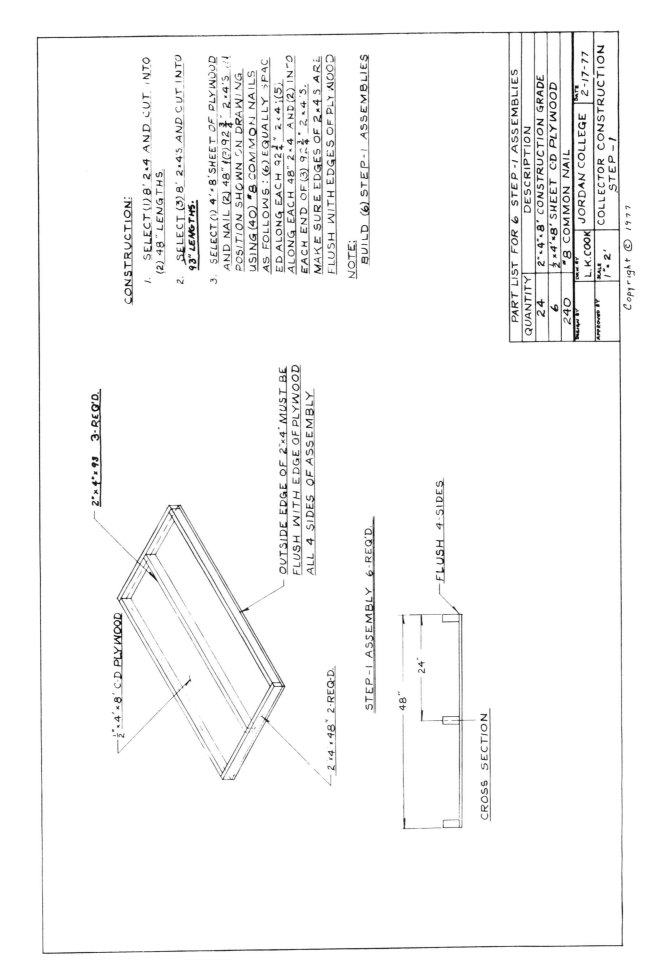

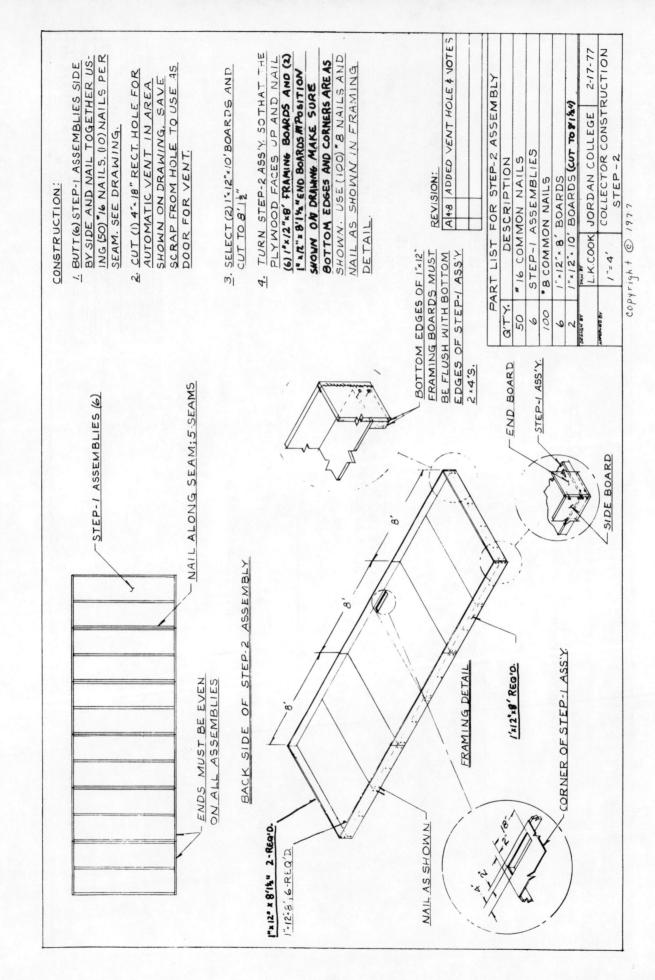

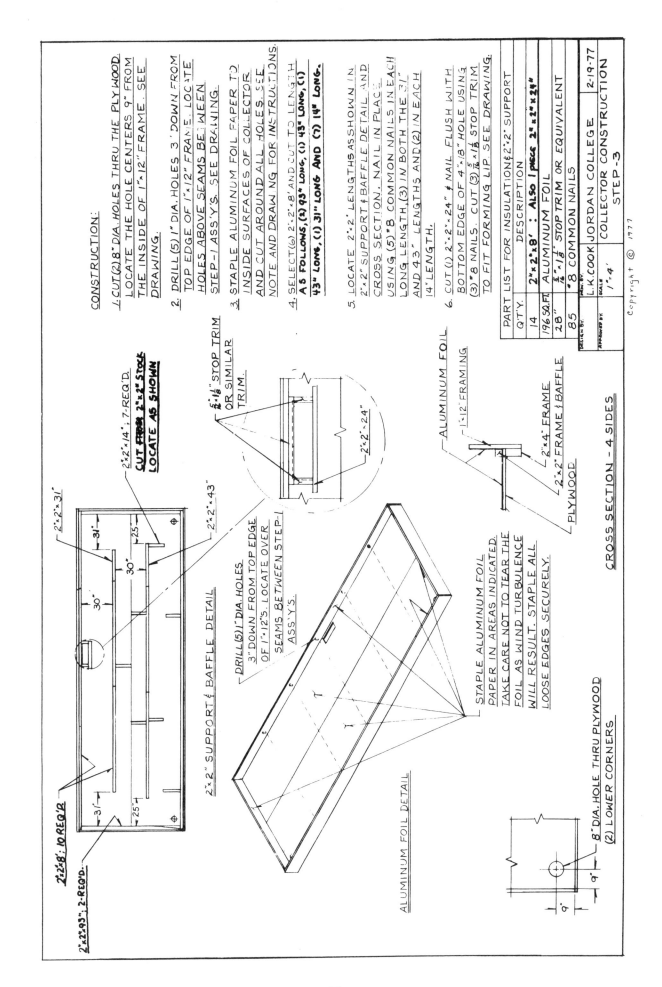

CONSTRUCTION:

1. CUT (1) 7" DIA. HOLE IN EACH OF (2) .016" x 36" x 93½" ALUMINUM SHEETS. LOCATE HOLE CENTERS 8" FROM EDGES OF THE SHEET. SEE DRAWING.

2. LOCATE ALUM. SHEETS AT ENDS OF COLLECTOR SO THAT THE 7" HOLES ARE ALIGNED OVER THE 3" HOLES IN THE COLLECTOR AND NAIL ON APPROXIMATELY 8" CENTERS. USE (21) 1½" ALUM. NAILS PER SHEET.

3. FOR OPTION #1 ADD (6) ADDITIONAL ALUM. SHEETS NAILED EVERY 8" ON BOTH ENDS AND ON (2) 2" x 2" SUPPORT.

4. FOR OPTIONS #2-#4 ADD (6) ALUM. SHEETS AS IN OPT #1 BUT DO NOT NAIL ALONG 2" x 2" SUPPORTS UNTIL AFTER ALUM. EXTRUSIONS ARE NAILED IN POSITION. USE 1½" ALUM. NAILS. SEE OPTION LIST.

5. FOR OPTION #5 NAIL EXTRUSIONS DIRECTLY TO 2" x 2" SUPPORTS USING 1½" ALUMINUM NAILS.

IMPORTANT:
4" x 18" HOLE FOR VENT MUST BE CUT IN ALUMINUM SHEET DIRECTLY OVER 4" x 18" VENT HOLE IN COLLECTOR. ALUMINUM EXTRUSIONS MUST BE CUT TO 90" FOR USE IN THIS AREA 4-SQ.

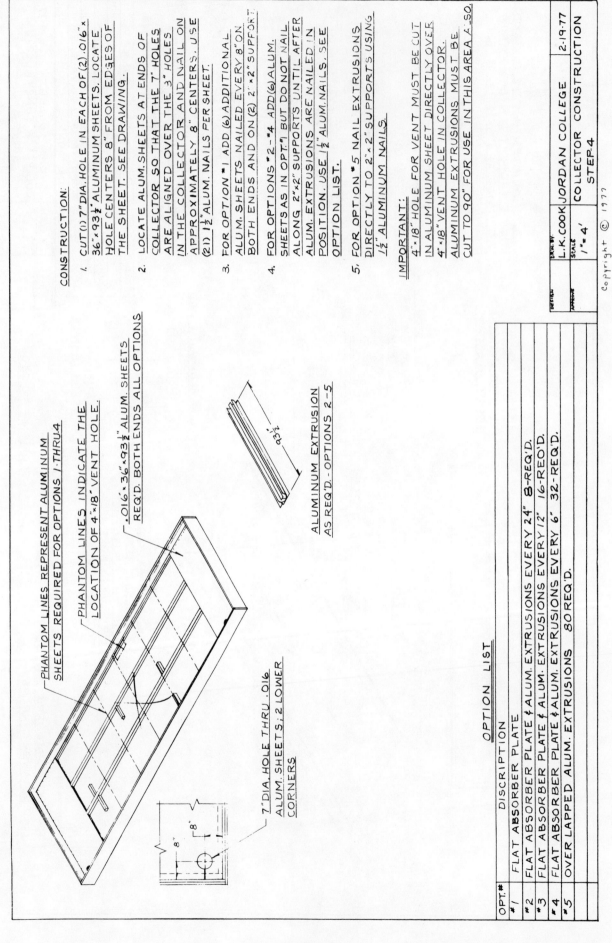

OPTION LIST

OPT.#	DISCRIPTION
#1	FLAT ABSORBER PLATE
#2	FLAT ABSORBER PLATE & ALUM. EXTRUSIONS EVERY 24" 8-REQ'D.
#3	FLAT ABSORBER PLATE & ALUM. EXTRUSIONS EVERY 12" 16-REQ'D.
#4	FLAT ABSORBER PLATE & ALUM. EXTRUSIONS EVERY 6" 32-REQ'D.
#5	OVER LAPPED ALUM. EXTRUSIONS 80 REQ'D.

DRWN BY: L.K. COOK | JORDAN COLLEGE | 2-19-77
SCALE 1"=4' | COLLECTOR CONSTRUCTION STEP-4

Copyright © 1977

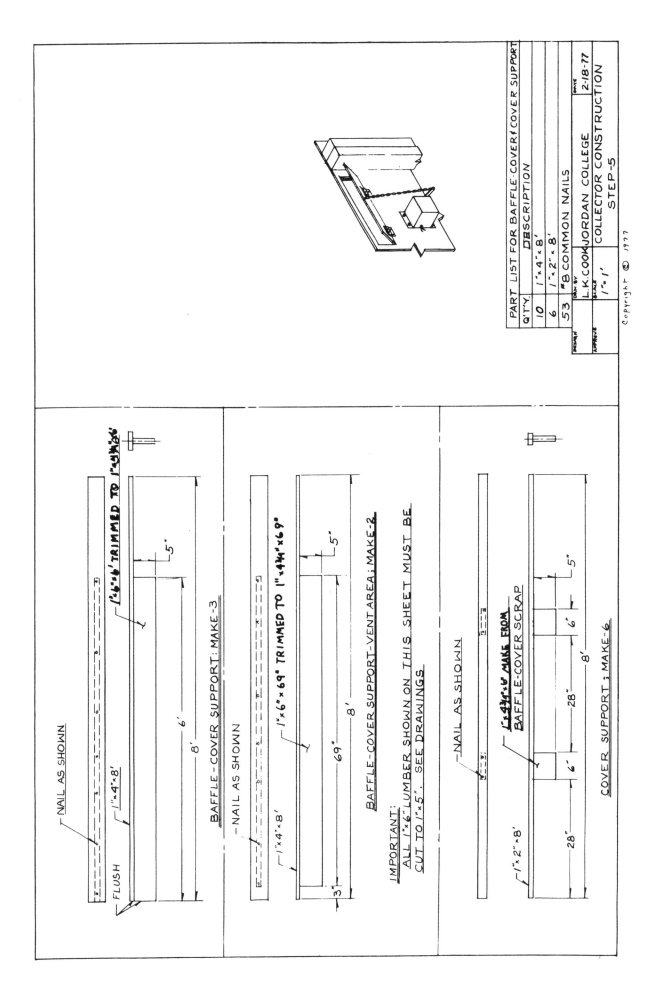

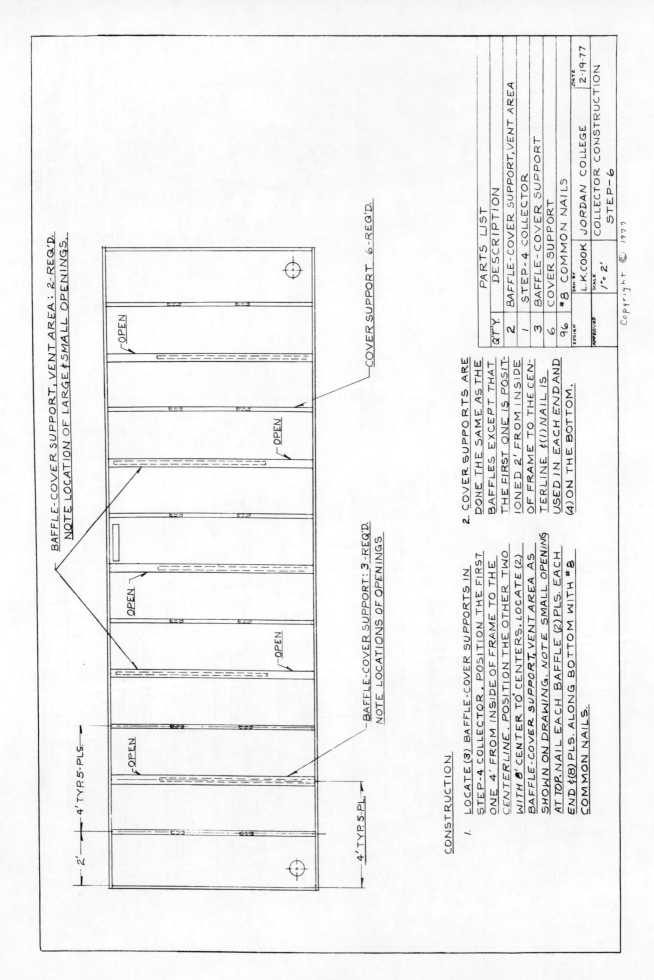

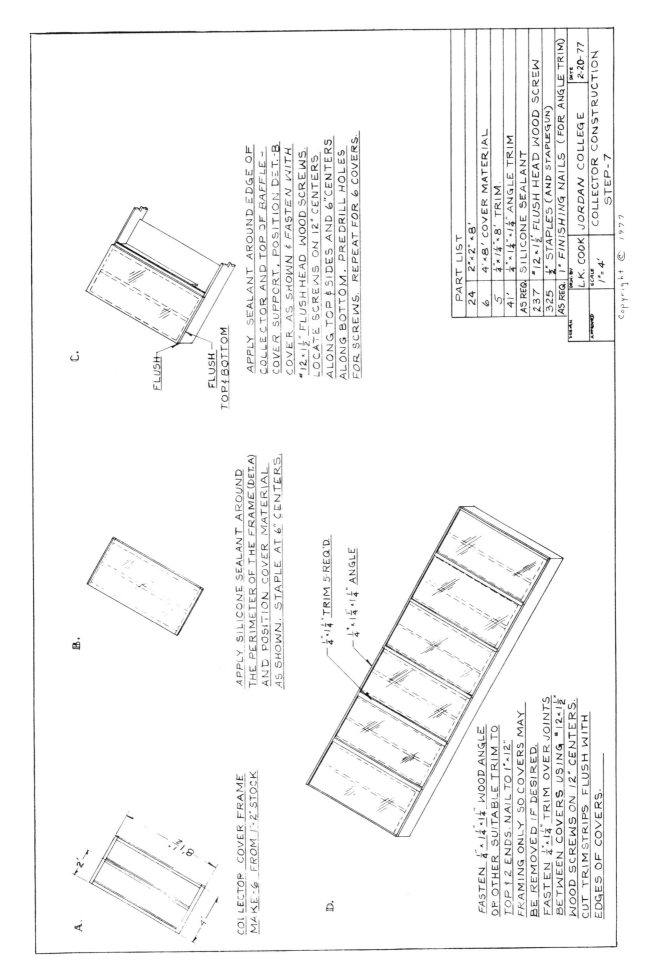

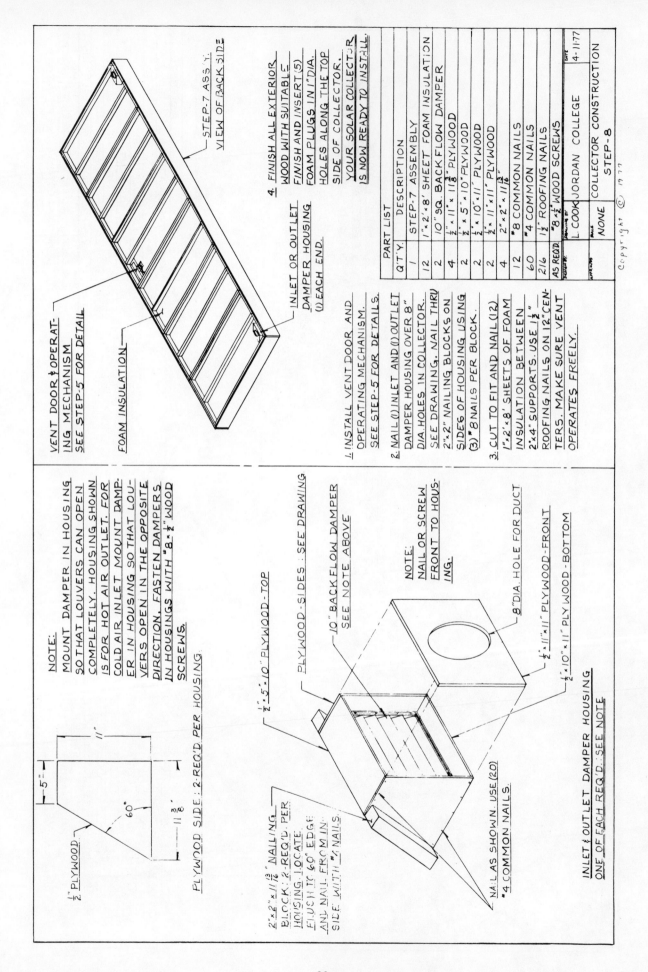

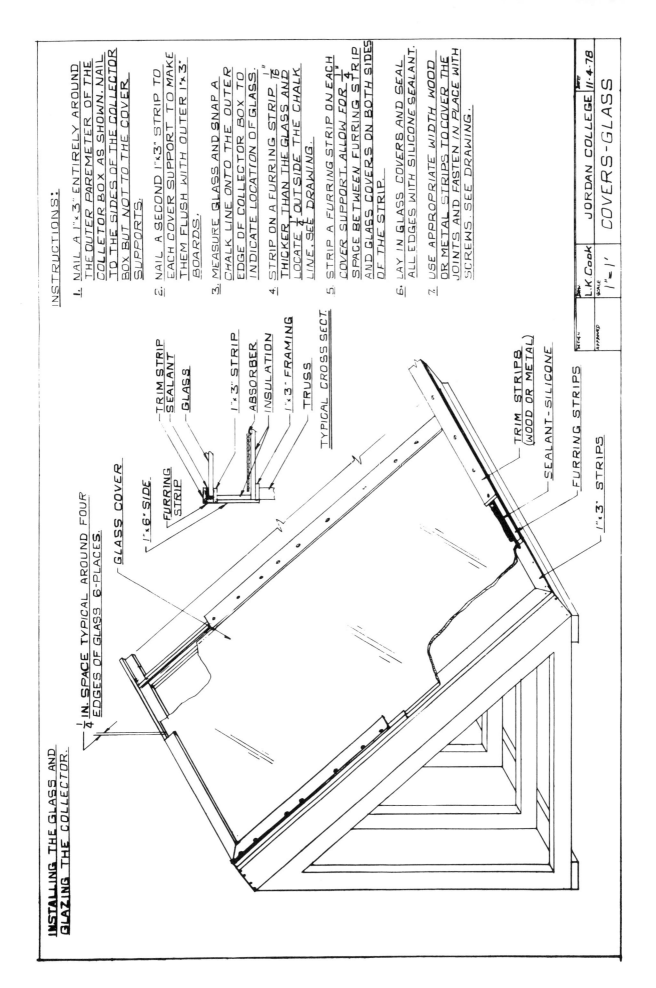

Manifold Style Collector

The Manifold Style designed at a 50° pitch is designed for adaptation to roofs better than the Serpentine due to the ease of building with trusses rather than 4 x 8 sections. Two drawings are given in order that the choice can be made between erecting the collector on the ground or on the roof. Drawing three (3) shows Detail A and B, explaining the difference in size if glass is used. The reason for the 2 x 4 kicker at the bottom of Detail B is to support the weight of the glass. Tests showed, without such support glass would cause the framing on Detail A to sag.

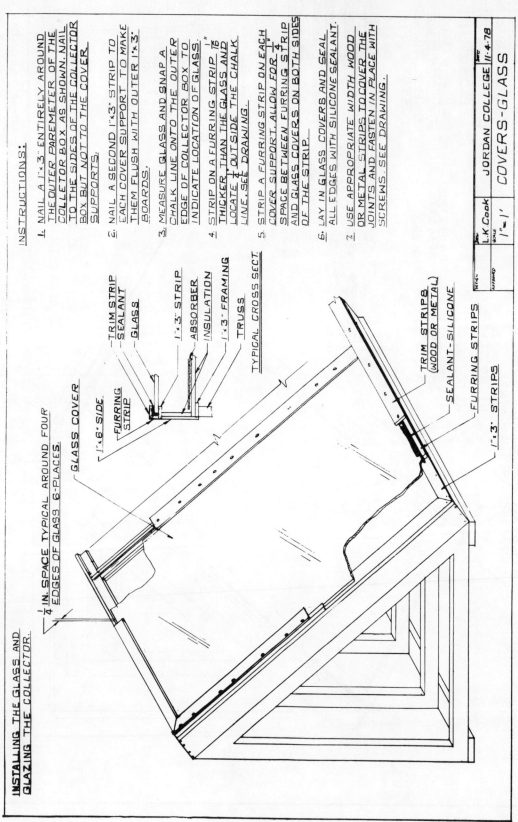

TYPES OF SOLAR BASES FOR TRUSSES TO BE MOUNTED AT 50°.

EXAMPLE 1: ON THE GROUND ~~WITH WATER HANDLER AND STORAGE IN THE BASEMENT.~~

INSTRUCTIONS:

1. DRIVE (4) STAKES IN THE GROUND TO FORM (2) 30-FT LONG PARALLEL LINES 4-FT APART.
2. POUR CEMENT FOOTINGS 42" DEEP, 8" WIDE AND 24-FT LONG AS SHOWN.
3. PLACE (16) ANCHOR BOLTS IN CONCRETE. LOCATE BOLTS ON 3'10" CENTERS. SEE DRAWING AND ANCHOR BOLT DETAIL.
4. LOCATE AND DRILL 5/8" HOLES FOR ANCHOR BOLTS IN (4) 2"X6"X12FT PLATES AND BOLT TO FOOTINGS AS SHOWN.
5. CALK OVER THE ANCHOR BOLTS FOR WEATHERIZATION.
6. PREFERRED MATERIAL FOR PLATES — "MERCHANTABLE GRADE" REDWOOD.

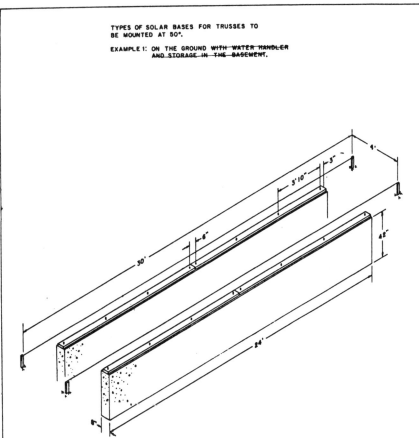

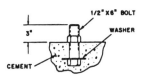

ANCHOR BOLT DETAIL

| DESIGN | DRW R. MILLER | JORDAN COLLEGE | DATE 8-6-79 |
| APPROVE | SCALE | BASE - GROUND INSTALLATION | 1 |

TYPES OF SOLAR BASES WITH TRUSSES TO BE MOUNTED ON ROOF.

EXAMPLE 2: ON A FLAT OR A PITCHED ROOF NOTE: FOR PITCHED ROOF DETERMINE THE ANGLE OF THE PITCH AND BUILD PLATFORM TO CARRY TRUSS ASSEMBLY AS SHOWN IN DRAWING NO. 4.

INSTRUCTIONS:

1. SNAP (2) PARALLEL CHALK LINES 28-FT LONG AND 4-FT APART.
2. CUT (14) 2"X4" SPACERS 12" LONG, LOCATE AND AND TACK TO ROOF ON 4-FT CENTERS.
3. TACK THE 2"X6" PLATES TO THE SPACERS.
4. DRILL 5/8" HOLES THROUGH PLATES, SPACERS AND ROOF AS SHOWN.
5. INSERT ANCHOR BOLTS IN HOLES, SLIP ROOF SUPPORTS IN PLACE AND TIGHTEN NUTS ON BOLTS. SEE ANCHOR BOLT DETAIL.
6. APPLY CALK TO EXTERIOR BOLTS OR NUTS.
7. PROTECT THE 2"X4" SPACERS BY APPLYING METAL FLASHING IN ACCORDANCE WITH PROPER BUILDING PROCEDURES.

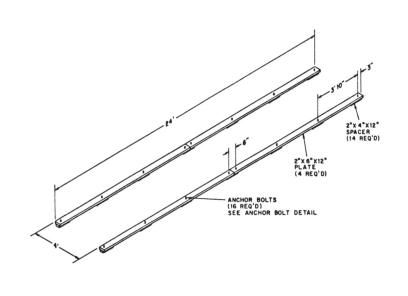

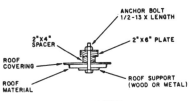

ANCHOR BOLT DETAIL

| DESIGN | DRW R MILLER | JORDAN COLLEGE | DATE 8-7-79 |
| APPROVE | SCALE | BASE - ROOF INSTALLATION | 2 |

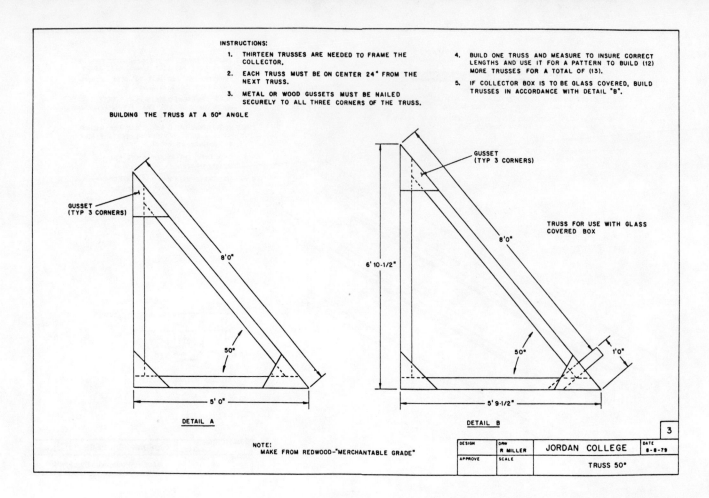

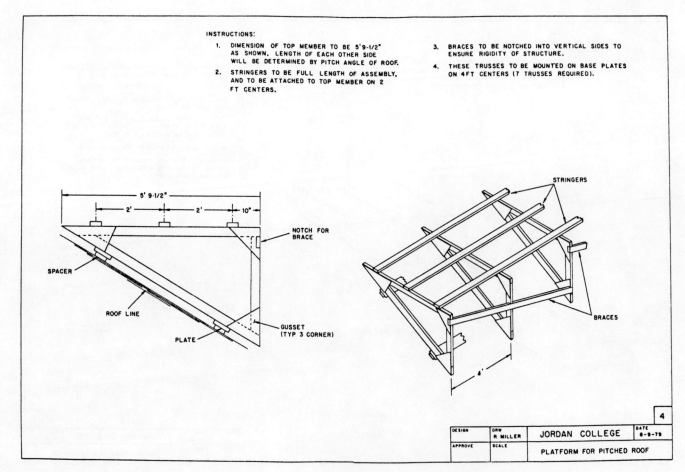

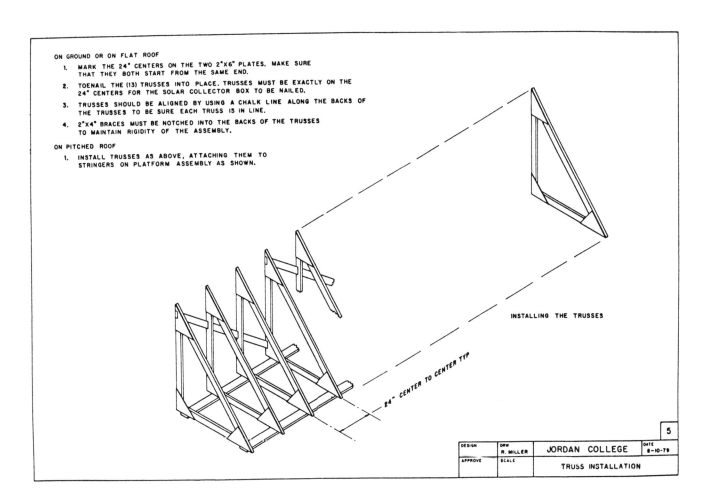

ON GROUND OR ON FLAT ROOF
1. MARK THE 24" CENTERS ON THE TWO 2"X6" PLATES. MAKE SURE THAT THEY BOTH START FROM THE SAME END.
2. TOENAIL THE (13) TRUSSES INTO PLACE. TRUSSES MUST BE EXACTLY ON THE 24" CENTERS FOR THE SOLAR COLLECTOR BOX TO BE NAILED.
3. TRUSSES SHOULD BE ALIGNED BY USING A CHALK LINE ALONG THE BACKS OF THE TRUSSES TO BE SURE EACH TRUSS IS IN LINE.
4. 2"X4" BRACES MUST BE NOTCHED INTO THE BACKS OF THE TRUSSES TO MAINTAIN RIGIDITY OF THE ASSEMBLY.

ON PITCHED ROOF
1. INSTALL TRUSSES AS ABOVE, ATTACHING THEM TO STRINGERS ON PLATFORM ASSEMBLY AS SHOWN.

INSTALLING THE TRUSSES

24" CENTER TO CENTER TYP

JORDAN COLLEGE — TRUSS INSTALLATION — DRW R. MILLER — DATE 8-10-79 — 5

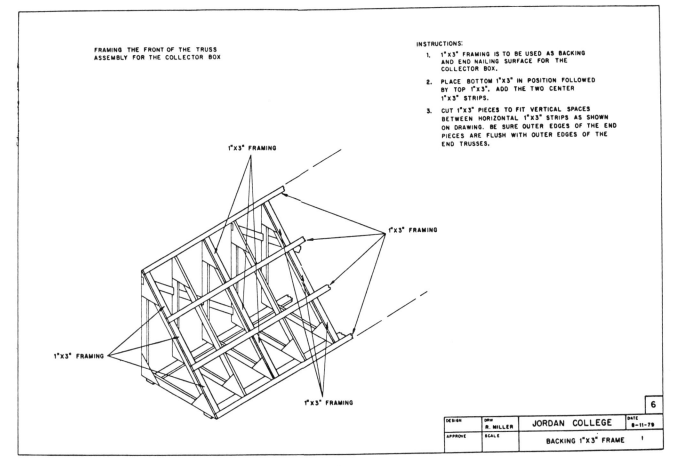

FRAMING THE FRONT OF THE TRUSS ASSEMBLY FOR THE COLLECTOR BOX

INSTRUCTIONS:
1. 1"X3" FRAMING IS TO BE USED AS BACKING AND END NAILING SURFACE FOR THE COLLECTOR BOX.
2. PLACE BOTTOM 1"X3" IN POSITION FOLLOWED BY TOP 1"X3". ADD THE TWO CENTER 1"X3" STRIPS.
3. CUT 1"X3" PIECES TO FIT VERTICAL SPACES BETWEEN HORIZONTAL 1"X3" STRIPS AS SHOWN ON DRAWING. BE SURE OUTER EDGES OF THE END PIECES ARE FLUSH WITH OUTER EDGES OF THE END TRUSSES.

1"X3" FRAMING

JORDAN COLLEGE — BACKING 1"X3" FRAME — DRW R. MILLER — DATE 8-11-79 — 6

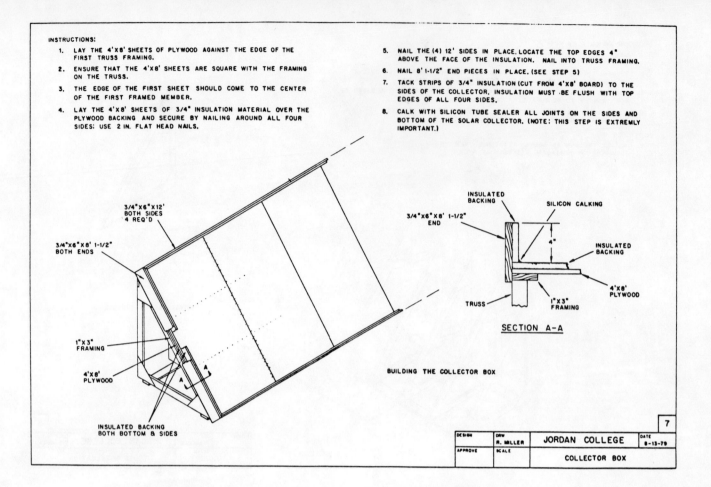

Ducting the air collector is done according to job site needs. Form openings at the top and form openings at the bottom of the collector insure adequate air flow. Openings should be approximately 4"x12 inches and regular HVAC components can be used. The duct work should be well insulated, preferrable from the inside. Air should be drawn out from the top of the collector and not pumped in.

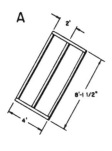

A

COLLECTOR COVER FRAME
MAKE (6) FROM 1"x2" STOCK.

B

APPLY SILICONE SEALANT AROUND
THE PERIMETER OF THE FRAME (DET A)
AND POSITION COVER MATERIAL
AS SHOWN. STAPLE AT 6" CENTERS.

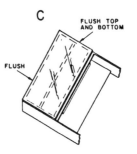

C

FLUSH TOP AND BOTTOM

FLUSH

APPLY SEALANT AROUND EDGE OF
COLLECTOR AND TOP OF BAFFLE-
COVER SUPPORT. POSITION DET B
COVER AS SHOWN AND FASTEN WITH
NO. 12 X 1 1/2" FLUSH HEAD WOOD SCREWS.
LOCATE SCREWS ON 12" CENTERS
ALONG TOP AND SIDES AND 6" CENTERS
ALONG BOTTOM. PREDRILL HOLES
FOR SCREWS. REPEAT FOR 6 COVERS.

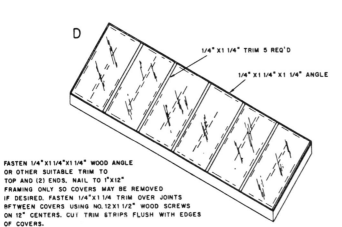

D

1/4" X 1 1/4" TRIM 5 REQ'D
1/4" X 1 1/4" X 1 1/4" ANGLE

FASTEN 1/4"X 1 1/4"X 1 1/4" WOOD ANGLE
OR OTHER SUITABLE TRIM TO
TOP AND (2) ENDS. NAIL TO 1"X12"
FRAMING ONLY SO COVERS MAY BE REMOVED
IF DESIRED. FASTEN 1/4"X 1 1/4 TRIM OVER JOINTS
BETWEEN COVERS USING NO. 12 X 1 1/2" WOOD SCREWS
ON 12" CENTERS. CUT TRIM STRIPS FLUSH WITH EDGES
OF COVERS.

DESIGN	DRW R. MILLER	JORDAN COLLEGE	DATE 8-21-79
APPROVE	SCALE	COLLECTION BOX COVERS PLASTIC	10

INSTALLING THE GLASS

INSTRUCTIONS:

1. NAIL A 1"X3" STRIP ALONG THE UPPER SIDE AND BOTH ENDS OF THE COLLECTOR BOX AS SHOWN. NAIL TO THE SIDES OF THE COLLECTOR BOX BUT NOT TO THE COVER SUPPORTS.

2. NAIL A SECOND 1"X3" STRIP TO EACH COVER SUPPORT TO MAKE THEM FLUSH WITH OUTER 1"X3" BOARDS.

3. MEASURE GLASS AND SNAP A CHALK LINE ONTO THE OUTER EDGE OF COLLECTOR BOX TO INDICATE LOCATION OF GLASS. (ALLOW FOR GLASS TO BE SEATED 1/4" IN GROOVE IN LOWER SIDE.)

4. STRIP ON A FURRING STRIP 1/16" THICKER THAN THE GLASS AND LOCATE 1/4" OUTSIDE THE CHALK LINE. SEE DRAWING.

5. STRIP A FURRING STRIP ON EACH COVER SUPPORT. ALLOW FOR 1/4" SPACE BETWEEN FURRING STRIP AND GLASS COVERS ON BOTH SIDES OF THE STRIP.

6. LAY IN GLASS COVERS AND SEAL ALL EDGES WITH SILICONE SEALANT

7. USE APPROPRIATE WIDTH WOOD OR METAL STRIPS TO COVER THE JOINTS AND FASTEN IN PLACE WITH SCREWS. SEE DRAWING.

8. INSULATION APPLIED TO LOWER SIDE TO BE 4-3/4" WIDE (FLUSH WITH INNER EDGE OF GLASS GROOVE).

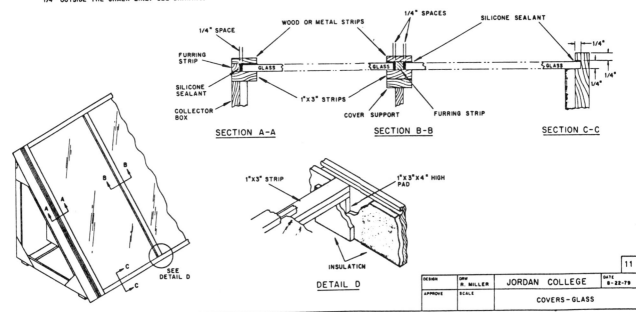

DESIGN	DRW R. MILLER	JORDAN COLLEGE	DATE 8-22-79
APPROVE	SCALE	COVERS-GLASS	11

AIR HANDLING

There are four basic modes of air handling. These modes are:
1. house—collector—house
2. collector—storage—collector
3. house—storage—house
4. mixing—collector & storage—house

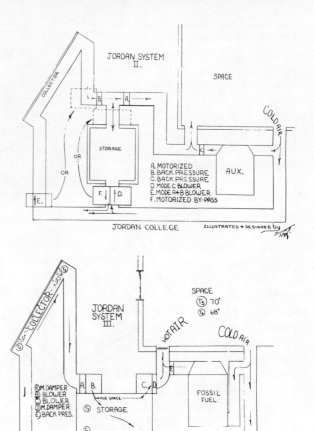

Mode I
House—Collector—House

You will notice that diagram one shows the air flow in mode one. In that print the air is drawn through the back pressure damper at the base of the storage, through the collector and into the cold air return of the furnace system. Solar air may be blown directly into the building without entering the furnace if so desired. Warmed air from the collector should never be stored if the thermostat calls for heat as BTU's are always lost in the heat transfer process. During continually cold winter days, storage may not take place as solar heat is immediately used when house temperatures demand it.

Mode II
Collector—Storage—Collector

The second mode shows the system in the storage mode. Air is simply drawn through the collector, cycled through the storage and back into the collector. The storage loop builds up heat as the air pre-heats the rocks. Eventually the fully charged rocks will store enough heat for three days. In some cases, as through the summer months, the system can be operated manually if charged storage is not desired. During this time the collector external exit for air will discharge collector heat.

Mode III
House—Storage—House

The heat that is taken from the storage mode is drawn in reverse flow through the storage into the house. The air handler diverts the storage air into the house by closing the damper into the collector. The back pressure damper at the other end of the collector will not open to allow air from the collector thus drawing the floor damper in the storage to pull open.

Mode IV
Mixing—Collector & Storage—House

Storage Considerations for Jordan Air Systems

Calculations for storage on 8 x 24 foot collectors as shown in the Serpentine and Manifold designs can be calculated as follows.

The 192 square feet of absorber will generate approximately 30,000 BTU's/hr. at 245/BTU/ft$_2$/hr. During the shortest winter days in December, January and February the sun gives about five hours of useable energy. The fall and spring has expanded sun lengths of seven to nine hours. Figuring the model on a typical fall or spring day, the collector would generate 30,000 BTU's/hr. for eight hours for a total of 240,000. The 4 x 4 x 12 foot rock storage bin will store, when fully charged, 270,000 BTU's on one full day of sun. But, being that in the fall and spring heat will be needed overnight and mornings, the charge can last into three days, while it charges and discharges through use.

Storage

The purpose of including storage details is to aid in retrofitting existing buildings. New structures generally will have buried stone storage or compartments natural to the building. Care should be taken that adequate insulation and support are part of such structures.

The drawings illustrate a typical storage bin

suitable for use in buildings located between 40 degree and 50 degree latitude. The suggested volume of rock storage material is adequate to provide up to three days' useful heat storage. Variations in the design will probably be necessary for some homeowners, depending upon the shape and position of available space. One simple rule must be remembered: hot air from the collector must go in at the top of the storage bin, and be drawn out at the bottom, then return to the collector. Also, the back pressure damper at the collector inlet must be drawn in.

Step I

Building the four box supports is the first step in the storage construction. Some may wish to add three 2 x 4 box supports if rock pressure becomes too great. White glue is recommended between 2" x 4" scrap used at the corners of the square box supports.

The measurement of pieces as described in the materials list on the lower right hand corner should be followed closely. Be sure the 1 x 6 x 59 piece is attached to the correct frames. Before nailing the plywood to the bottom, look at the position of the end frames as to correct position of the 2 x 6 x 59 pieces. Plywood can now be slid through the frames and nailed at the appropriate distances. The 4 x 8 foot piece should be placed at the middle of the 2 x 6 that butts against the 4 x 4 foot piece of plywood.

Step II

Two assemblies are shown in the second step. The end sections should be cut as described in the materials list. They are simply constructed as the measurements show.

The assembly of the sides, grates, and top should be preceded by careful study of prints. The plywood sides must be nailed to the framing before the bricks are positioned over the framing. The grates should be of such quality metal as to sustain the tonnage in the storage unit. Remember that the four-foot span will demand heavy metal. The air space created by the bricks is critical to the removal of warm air coming through the stones. The air space must continue through the end that houses the back pressure damper and cold air return to the collector. Place the copper tube in position.

After nailing the two plywood ends into place filling of the bin can begin. The necessary insulation and finishing of all parts should be completed before filling. Stones should be poured carefully. If bulging takes place, bracing should be used. Stones should only be filled to within eight inches of the top. Do not overfill. The air space is necessary for hot air movement through the stones.

Step III

The materials list should be reviewed as the instructions in notes 1 through 5 are important to consider before proceeding. Be sure and cut the air ducts into the 4 x 8 plywood sheet used as the top of the collector before nailing it into place. You may wish to slide the top plywood into place and then lay the exterior cover stock over it. Then drill through the top material and cut that hole. Use the hole thus formed as a template to draw the position of the corresponding hole in the plywood. Remove the outer cover, cut the hole in the 4 x 8 plywood, and secure this piece by nailing as indicated.

Now install the six-inch fiberglass insulation blanket as shown in the drawing, and attach the cover. Do **not** insulate the space at the bottom of the cold-air duct end of the bin (see note on drawing). Install the 10-inch back-flow damper, taking care that air can enter but not exit the storage.

Complete the installation by inserting the sensor into the copper tubing, and connecting the storage by suitable ducting to the collector and air handling unit.

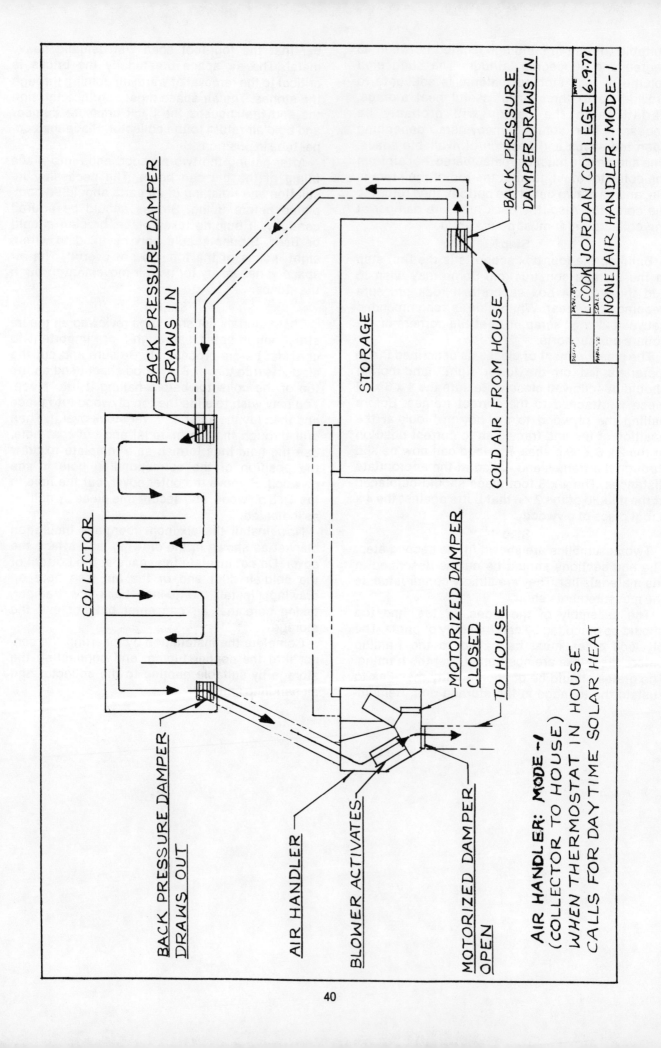

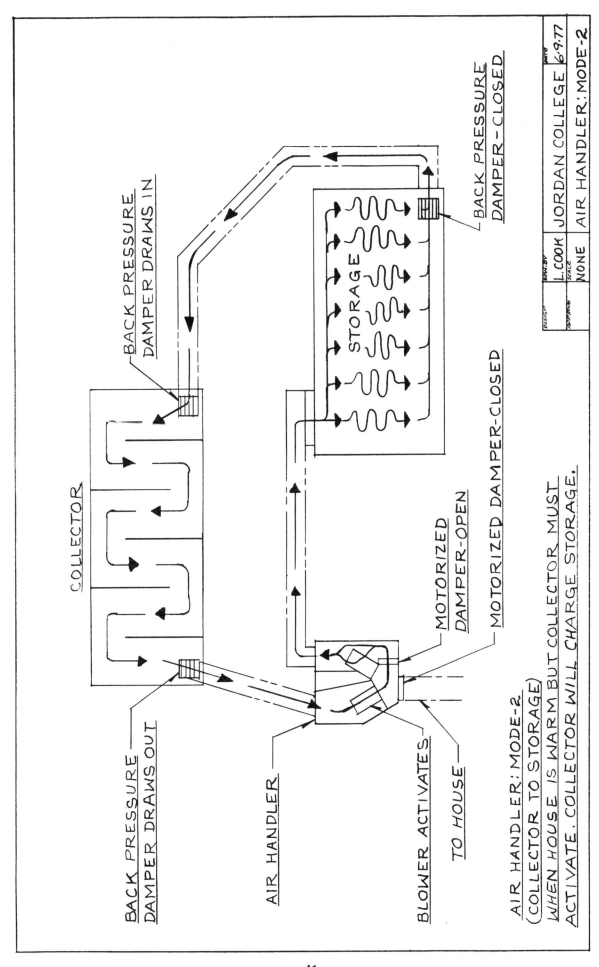

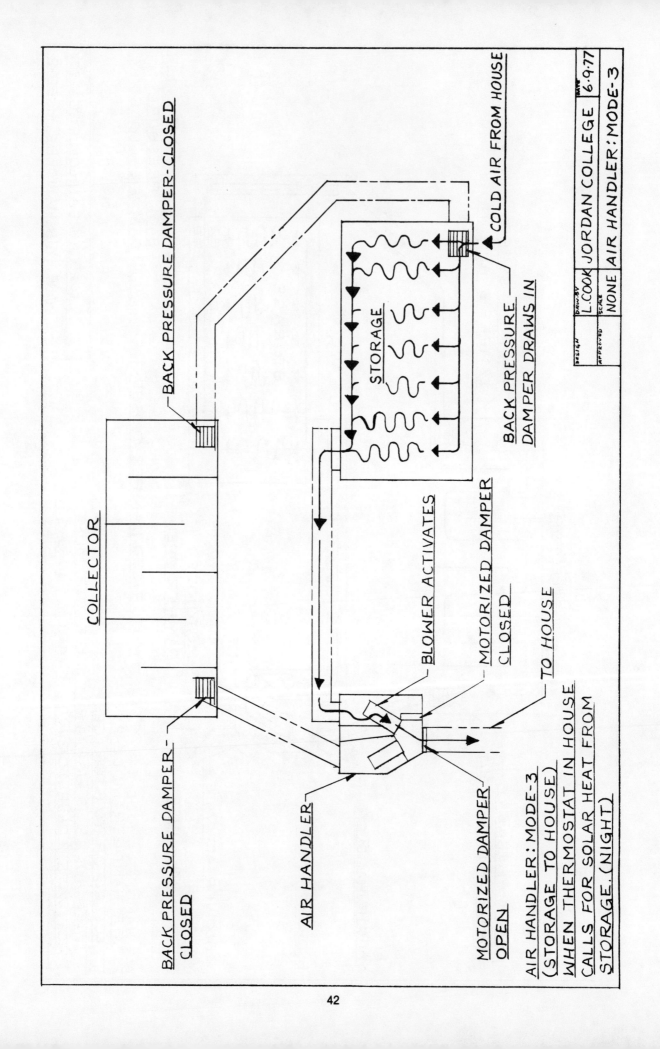

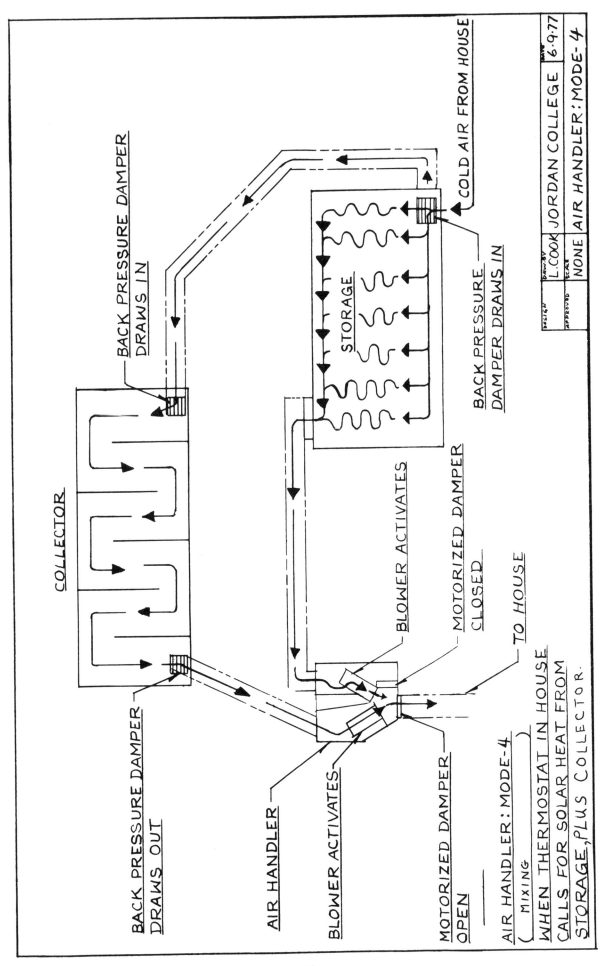

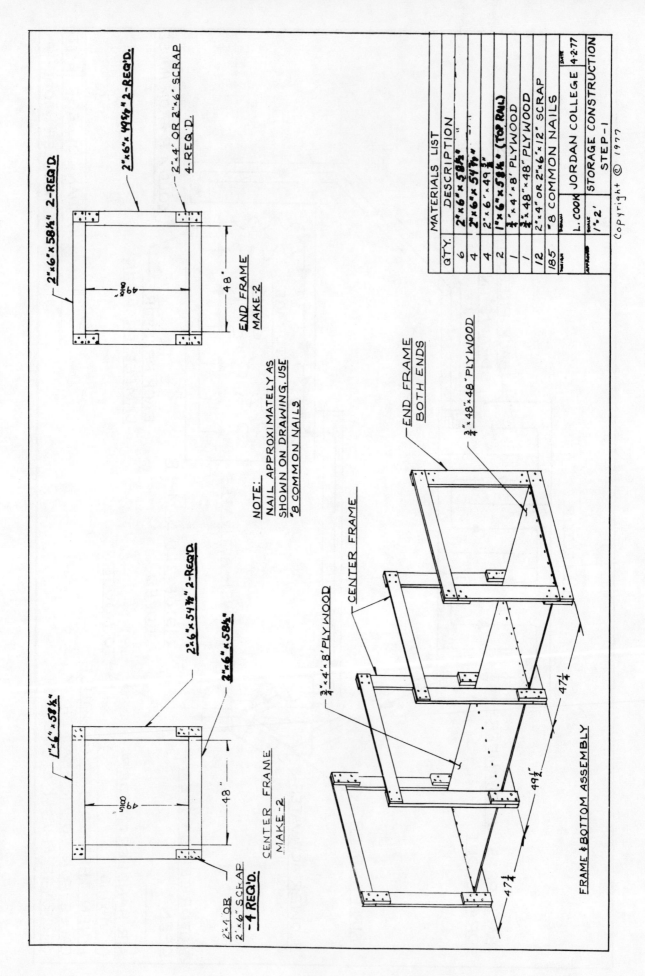

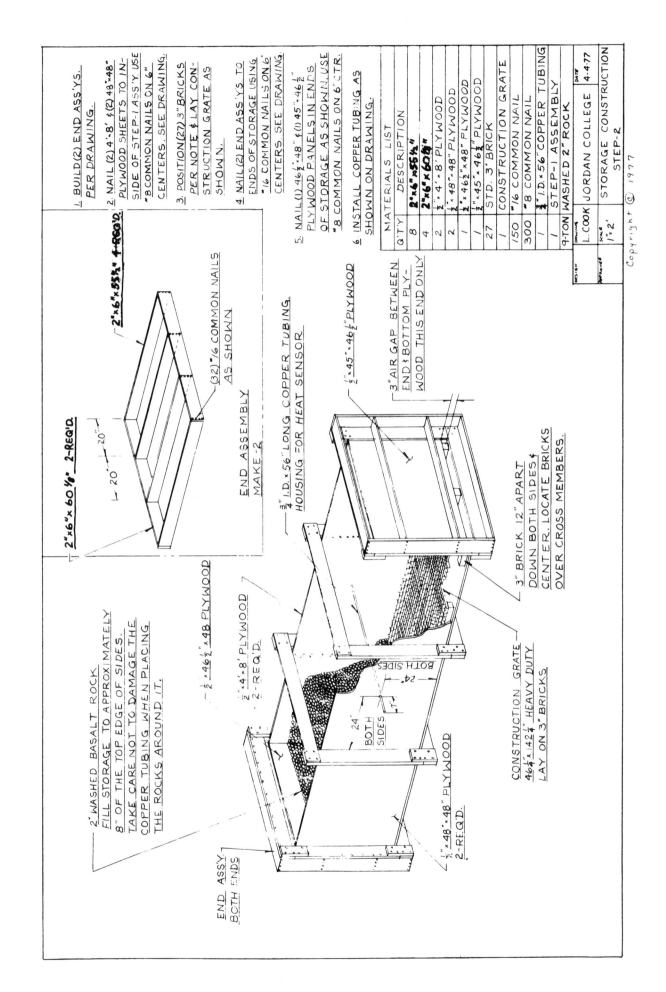

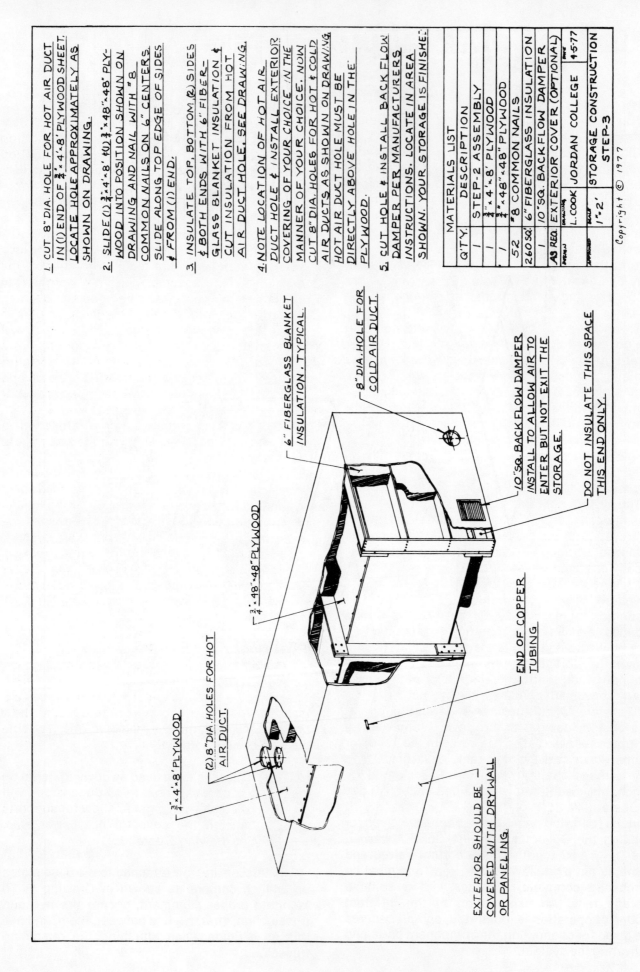

1. CUT 8" DIA. HOLE FOR HOT AIR DUCT IN (1) END OF 3/4"x4'x8' PLYWOOD SHEET. LOCATE HOLE APPROXIMATELY AS SHOWN ON DRAWING.

2. SLIDE (1) 3/4"x4'x8' & (2) 3/4"x48"x48" PLYWOOD INTO POSITION SHOWN ON DRAWING AND NAIL WITH #8 COMMON NAILS ON 6" CENTERS. SLIDE ALONG TOP EDGE OF SIDES & FROM (1) END.

3. INSULATE TOP, BOTTOM, (2) SIDES & BOTH ENDS WITH 6" FIBERGLASS BLANKET INSULATION & CUT INSULATION FROM HOT AIR DUCT HOLE. SEE DRAWING.

4. NOTE LOCATION OF HOT AIR DUCT HOLE & INSTALL EXTERIOR COVERING OF YOUR CHOICE IN THE MANNER OF YOUR CHOICE. NOW CUT 8" DIA. HOLES FOR HOT & COLD AIR DUCTS AS SHOWN ON DRAWING. HOT AIR DUCT HOLE MUST BE DIRECTLY ABOVE HOLE IN THE PLYWOOD.

5. CUT HOLE & INSTALL BACKFLOW DAMPER PER MANUFACTURERS INSTRUCTIONS. LOCATE IN AREA SHOWN. YOUR STORAGE IS FINISHED.

MATERIALS LIST

QTY.	DESCRIPTION
1	STEP-2 ASSEMBLY
1	3/4"x4'x8' PLYWOOD
2	3/4"x48"x48" PLYWOOD
52	#8 COMMON NAILS
260 SQ.	6" FIBERGLASS INSULATION
1	10" SQ. BACKFLOW DAMPER
AS REQ.	EXTERIOR COVER (OPTIONAL)

DRAWN BY: L. COOK JORDAN COLLEGE DATE: 1-5-77
SCALE: 1"=2'
STORAGE CONSTRUCTION STEP-3

Copyright © 1977

Jordan Solar Liquid System

General Description

The Jordan "drain-down" Liquid Solar System is a simplified system using solar energy to heat a liquid. This liquid is contained in a closed circuit comprising a storage tank, a "day tank" (which serves a purpose similar to a domestic boiler) and a solar energy collector. A heat exchanger in the day tank provides heated water for domestic use.

Inclusion of the heat exchanger makes it possible to use either water or an anti-freeze mixture in the closed circuit; the anti-freeze feature may be considered desirable in any areas where below freezing temperatures are frequently experienced.

The drain-down principle allows the liquid in the collector to drain by gravity into the day tank whenever conditions make this necessary, thus avoiding the possibility of freezing in the collector. As the liquid drains down, it is replaced by air (contained in the top of the day tank). In order for this to occur, the collector must be elevated above the day and storage tanks. If the tanks can be installed in a basement, the collector may be sited at ground level; otherwise, it will be necessary to locate the collector on the roof.

The system operates in three modes, the selection of each mode being fully automatic. The Jordan system is unique in that mode selection is achieved using only one pump and four electrically operated valves. All necessary circuitry is contained on one circuit board, and the entire assembly of pump, valves and circuit board is mounted on the day tank. External wiring is thus reduced to a minimum, requiring only connection to a power source and leads to temperature sensors in the storage tank and collector box, and to a temperature controller which may be mounted at any convenient point near the day tank.

"Fail-safe" operation is an important feature of this system design. If the power supply is interrupted for any reason the liquid in the collector drains down into the day tank, thus preventing any damage through freezing. The method by which this is achieved is explained under "Modes of Operation".

In the following pages instructions are given for building the Collector Box and Solar Absorber. The Day Tank, together with control valves and pump, is not a "do-it-yourself" item, and must be purchased complete, together with a suitable storage tank. An explanation of the different modes of operation is illustrated, and procedures are given for connecting the component parts and activating the system.

BUILDING THE SOLAR COLLECTOR

In the interests of economy and ease of construction, the Collector Box and its supporting structure are built as one unit. Construction details will vary, depending on whether the collector is to be sited on the ground or on a flat or pitched roof.

Base: Details of the base for the collector box and its supporting trusses are shown in Drawings 1 and 2. Drawing 1 shows the construction of concrete footings which are required to support the truss and box assembly, which weighs approximately 1,800 pounds when glass is used as the cover material. Where fiberglass or plastic panels are used, total weight is reduced to approximately 1,200 pounds. Deep footings are provided for necessary stability in northern latitudes where freezing and thawing are common in winter. The base plates are secured by anchor bolts set in the concrete. After the plates are installed, caulk or other suitable preservative should be applied to the anchor bolts and nuts to prevent corrosion.

The base for roof installation is shown in Drawing 2. It is essential to ensure that the roof is strong enough to accept the weight of the collector structure, and to withstand any stresses imposed by the effect upon it of strong wind gusts.

To avoid the possibility of water leaking through the roof around the anchor bolts, proper flashing must be applied to protect each 2 x 4 spacer. It is strongly recommended that the advice of a builder be sought in this respect.

Truss Construction and Assembly

Drawing 3, detail A shows Truss construction for installation on the ground or on a flat roof. Care must be taken to ensure that all trusses are identical for proper support of the collector box, and that they are rigidly assembled, with suitable gussets at each corner.

NOTE: If glass is to be used as cover material on the collector box, truss dimensions will be somewhat larger to allow for supports beneath the collector box lower side. See drawing 3, detail B.

The trusses are now attached to the base plates on 24-inch centers as shown in Drawing 5. To facilitate proper alignment, secure the two end trusses first; stretch a line between them, and use this as a guide when attaching the remaining trusses.

Diagonal braces must now be attached to the back of the trusses to ensure that they remain upright. The importance of maintaining the rigidity of this assembly cannot be overemphasized; to this end, the truss members should be notched to accept the braces.

If the collector is to be installed on a pitched roof, the pitch angle of the roof must be determined and a platform constructed and mounted on the base plates as shown in illustration 4. The dimension of the platform members must be such that a level surface is produced, on which the collector box truss assembly can be mounted.

Collector Box Backing

The frame on which the collector box will be built comprises four horizontal members, with vertical filler strips, as shown in Drawing 6. Care must be taken to attach the perimeter strips so that their outer edges are flush with the outer edges of the end trusses.

Collector Box

The collector box is built up on the frame previously attached to the trusses. Six 4 x 8 sheets of ½-inch exterior grade plywood are applied as shown in Drawing 7; the exterior face of the ply should be primed and painted (or treated with wood preservative) before installation.

When applying the first sheet, care must be taken to align the outer edge of the ply flush with the corresponding outer edge of the framing members; the outer edge of the last sheet should then be flush with the edge of the framing. If it is not, the edges must be trimmed so that this requirement is met.

The 4 x 8 sheets of ¾-inch insulating material are now applied over the plywood and secured in a similar manner. Each sheet should be firmly attached by nailing along the edges and into the cross members of the framing.

Lengths of 1 x 8 wood trimmed to actual width of 6 inches are next applied by nailing to the framing, to form the sides and ends of the box. They must be so positioned that the top edges are 4" above the surface of the insulation backing.

NOTE: If it is proposed to use glass as the cover material, a different method of making and supporting the lower side of the box will be necessary, due to the weight of the glass. Details of this alternative construction are given in a later paragraph, and are illustrated in Drawing 11.

The sides and ends of the box are now lined with 4" wide strips of insulating material, cut from a 4 x 8 sheet. There must be no gaps where these strips meet, nor where they abut the main backing sheets. They must also be flush with the top edges of the sides and ends of the box.

This stage of construction is then completed by applying silicone sealant to all joints on the bottom and sides of the box, including the joints between each of the backing sheets. It is of the utmost importance that this step be carefully executed.

FABRICATING AND INSTALLING THE ABSORBER

The absorber is built up using ½-inch "D" section copper tubes fitted into one-inch copper pipes which serve as upper and lower header chambers. These one-inch pipes are supplied with the "D" shaped holes already stamped out. They should be cut into four-foot lengths with the first hole three inches from the end of the pipe. This permits couplers to be slipped over the ends of the header pipes to join two or more sections together.

It is strongly recommended that silver solder be used for all soldering operations on the absorber assembly. It produces a much stronger joint than the commonly used tinman's solder, and so reduces the likelihood of breaking a joint if the assembly should be mishandled.

The D-shaped tubing is cut into six-foot lengths, eight of which are required for each absorber section, together with two four-foot lengths of one-inch header tube. It is important that the "D" tubes should not project too far into the headers, since this might prevent proper draining of the absorber units, resulting in freezing and damage to the absorber during very cold weather. Therefore, before commencing to solder the "D" tubes into the headers, slide a length of ¾-inch dowel into the latter; this will allow the "D" tubes sufficient insertion to produce a firm soldered joint, while preventing them from causing any obstruction in the header chamber.

It is essential that in each section the "D" tubes are perpendicular to the upper and lower headers; also the headers must be parallel to each other and exactly the same distance apart.

To ensure that these requirements are met, and that all sections are identical, a simple jig should be devised, upon which the parts of each section can be assembled before soldering.

After eight "D" section tubes have been soldered into their upper and lower headers, the sheet copper absorber strips are next soldered to the flat side of the "D" section tubes. Center each tube on a six-inch sheet of copper strip and solder the strip to the tube. Great care must be taken that no gaps be allowed between the sheet copper and the tube; the two must be firmly soldered together through their entire length. The tube should be pressed tightly against the copper strip (using a heavy weight or some other suitable means) during the soldering operation; the formation of a proper soldered joint may be further assisted by coating the flat surface of the "D" section tube with liquid solder.

A total of five four-foot sections plus one section with headers three feet long will be required. The overall length of the absorber will thus be twenty-three feet; this allows for tilting the absorber in the collector box as described below, and provides clearance for the input and output pipe unions.

When the required number of sections have been completed, they should be placed in the Collector Box with the "D" tubes facing the front of the box. The headers should then be joined together by slip couplings which are soldered in place over the open ends of each header chamber. The end of the lower header on the last section is closed with a cap, as is the end of the upper header on the first section.

The entire assembly is then arranged in the Collector Box at an angle of 3°-5° (as shown in Drawing 8) so that, when erected, it will cause the liquid to drain down by gravity to the Day Tank.

The inlet and outlet pipes are then soldered in place; the inlet pipe to the open end of the lower header in the first section, and the outlet pipe to the last section of the upper header. The outlet pipe is run diagonally across the face of the absorber inside the Collector Box, as shown in Drawing 8. By keeping as much of the inlet and outlet pipes as possible inside the box, heat loss to the outside air is avoided without incurring the expense of insulating the pipes. Where the pipes are unavoidably exposed, they should be encased in Armaflex (or equivalent) insulation. This insulation should be painted to prevent deterioration from weather and the sun's ultraviolet rays.

The complete absorber assembly should now be pressure-tested to determine that no leaks are present. Cap one end of the tubing (either outlet or inlet) with a pressure gauge, and connect an air compressor to the other end. Establish a pressure of 50 PSIG, close the inlet valve and watch the gauge for any drop in the reading. If no drop in pressure is indicated after one hour, the absorber may be considered free of any leaks.

If leaks are present, (as shown by a drop in the pressure gauge reading) it will be necessary to check each soldered joint between the "D" tubes and the headers, the end caps, and all adapters and couplings. This is easily accomplished by brushing each joint with soapy water; bubbles will form at any leak.

When all joints have been tested, and any leaks located, release the pressure from the system and carefully re-solder each defective joint. Repeat the pressure test until no leaking is indicated.

The absorber should now be painted matt black over its entire outer surface, including the inlet and outlet pipes (where they run within the collector box), upper and lower headers, all "D" tubes and fins.

Cover Supports and Covers

The covering material, which may be glass or plastic, will be supported by 1" x 3" "T" section bearers positioned as shown in Drawing 9. The number of bearers and their spacing will be governed by the width of each panel of cover material. If 8' x 4' panels of glass or plastic are used, five bearers will be necessary, spaced on four-foot centers.

The bearers are made the full length of the box ends, and the sides of the box (and the side insulation) are notched so that the bearers are flush with the edges of the box sides. The web of each bearer must be notched to accommodate the diagonal outlet pipe. The bearers are secured by nailing or screwing at the ends into the box sides, and by nailing at four points along their length. At these positions, spacers will be required between the web of the bearer and the insulation backing; holes must be cut in the sheet metal of the absorber large enough to clear the spacers, so that these do not prevent expansion of the copper. Six-inch nails will be required at these positions; they will pass through the insulation backing, the supporting framing, and will penetrate the truss members, thus providing rigid support for the glass or plastic covers.

Plastic Covers

Plastic material such as fiberglass or "Tedlar" should first be attached to frames built from 1 x 2 inch stock as shown in Drawing 10. Note that each frame is coated around its perimeter with silicone sealant before the plastic sheet is applied. The sheeting is then secured by stapling all around the frame at six-inch centers.

When six of these units have been constructed, they are attached to the collector. Silicone sealant is first applied to the edges of the collector box sides and ends, and to the intermediate bearers. The frames are then set in place and fastened to the box using 1½-inch flat-heat wood screws, the screw heads being countersunk. Care should be taken that the end units are flush with the collector box ends; this will result in a gap of approximately ¼-inch between the frames where they rest in the bearers.

When the six frames have been secured, the edges are covered with trim strips, as shown in Drawing 10. The strips which cover the gaps between adjacent frames are secured using 1½-inch wood screws passing between the frames and securing the trim strips to the bearers. The angle trim which covers the sides and ends of the box is fastened to the frames with 1-inch finishing nails. This facilitates detaching the trim from the frames should any of the latter require removal from the Collector Box.

Glass Covers

If glass is to be used as the cover material, its weight necessitates a different method of construction of the collector box and the supporting trusses.

Each truss will include a length of 2" x 4" material projecting from the front truss member and perpendicular to it, to support the lower side of the collector box. In order that the truss shall measure a full 8 feet above this support, all truss dimensions will be somewhat increased; details are shown on Drawing 3, detail B.

The lower side of the collector box is 1½-inches wider than the other three sides, and is grooved to accept the glass. This side can be made from standard 1" x 8" lumber, which has an actual width of 7¼-inches, grooved as shown in Drawing 11.

Because of the necessity of maintaining the glass groove intact, the five intermediate bearers cannot be notched into the lower edge of the box. Instead, they are supported by a 4-inch length of 1" x 3" lumber, as shown in detail "D" of Drawing 11. This supporting pad should be secured to the collector box before the insulation backing is fastened in place.

It is important to note that the actual width of glass panels must be 47 inches (not 48 inches). This is to allow for the furring strips between each panel, and for expansion of the glass.

Dimensions and full instructions for applying the glass are given on the drawing. Note that each sheet of glass must be carefully sealed using a silicone sealant around each edge of the glass and in the groove in the lower side. It is of utmost importance that this sealing procedure be properly performed, so that no heat is lost by air entering or leaving the box around the edges of any panel.

STORAGE CONSIDERATION FOR THE JORDAN LIQUID SYSTEMS

One gallon of water will store 74 BTU's. The Jordan day tank is a 68 gallon unit that handles the liquid from the collector to house or storage. It will store 5,032 BTU and the rest must be stored in the size tank designed for the collectors installed. The 192 foot Jordan collector can generate a maximum of 30,000 BTU/hr. Multiplying this to a ten hour sun day a total of 300,000 BTU's could be generated in a day. Given that occupancy of the home demand zero use, one million BTU's could be generated in a three-four day period. Considering the equilibrium between collector and storage, the 192 square feet of collector should not go beyond one million BTU's collection. Thus,

$$\frac{1,000,000 \text{ BTU}}{74 \text{ BTU/gal}} = \text{a 13,500 gallon tank is needed}$$

for a storage unit for 192 square feet.

MODES OF OPERATION

MODE A   Day Tank—Collector—Day Tank

This is the primary mode of operation and occurs whenever the temperature in the collector (T_1) is higher than the temperature at the bottom of the day tank (T_2). This could happen, for example, when the collector receives morning sunlight after draining down overnight.

Valves 1 and 3 are open, 2 and 4 are closed. The pump activates, and the liquid is circulated from the bottom of the day tank, through the collector and is returned to the top of the day tank.

A manually-operated temperature controller (usually located near the day tank) is connected to a sensor in the upper part of the day tank. This controller will generally be set at a point between 150°F and 190°F. When the temperature of the liquid in the upper part of the tank (T_3) exceeds the setting on the controller (referred to as the "set-point"), while the collector temperature (T_1) is still greater than T_2, the system operation changes to Mode B.

MODE B
Day Tank—Collector—Storage Tank—Day Tank

The purpose of this mode is to store heat energy, and it will occur whenever there is little or no demand for domestic hot water and the collector continues to be heated by the sun's rays.

In this mode, valves 2 and 3 close and 4 opens; 1 remains open as before, and the pump continues to operate.

The liquid is now drawn from the bottom of the day tank, circulates through the collector to the bottom of the storage tank and flows from the top of the storage tank to the day tank.

If any demand on the system causes cold water to flow through the heat exchanger to an extent that lowers the temperature (T_3) of the liquid in the upper part of the day tank below the set-point, the system automatically reverts to Mode A. It remains in that mode until the temperature at the top of the day tank once more exceeds the set-point. The system then changes back again to Mode B.

MODE C Day Tank—Storage Tank—Day Tank

This mode is used when it is required to extract heat from the storage tank while no heat is being produced in the collector. This condition would occur, for example, after sunset, or during a period of prolonged overcast.

Conditions necessary to cause the system to enter Mode C are: (a) the temperature in the collector must be less than that in the bottom of the day tank (T_2), and (b) the temperature of the liquid in the storage tank (T_4) must be higher than that existing in the top of the day tank (T_3). It follows that this mode must succeed Mode B, since it is this latter mode which causes the increase in temperature in the storage tank.

Before the system enters Mode C, however, it is necessary that the liquid in the Collector shall drain down into the day tank. To effect this, Valve 3 opens, and the pump motor is temporarily de-energized. The liquid in the collector now drains down through the pump into the day tank, and is replaced by air which rises from the top of the day tank through Valve 3.

After the appropriate time has elapsed (pre-set at a period of between four minutes and seven minutes), Valves 1 and 4 close, and the pump is restarted. The liquid now circulates between the storage tank and the day tank until the temperature of the liquid in both tanks is the same.

At this point the pump ceases to operate and the system remains quiescent until either

    (a) the temperature in the day tank is reduced because of cold water entering the heat exchanger, or

    (b) the temperature in the collector rises above that at the bottom of the day tank.

In case (a), the pump is activated and the system continues to operate in Mode C until temperature equilibrium is restored. If the temperature difference between the storage tank and day tank falls below a previously determined and pre-set figure, which will usually be between 10°F and 20°F, the pump ceases to operate.

In case (b), where the temperature in the controller rises above T_2 (temperature at the bottom of the day tank), valve 2 closes, valve 1 opens, the pump is activated and the system operates in Mode A.

Fail-Safe Mode

In the event of a power failure valves one and three open (they are "normally-open" valves) and valves two and four close (these are "normally-closed" valves). The pump ceases to operate and the liquid in the collector drains down into the day tank. When power is restored, the system automatically returns to whichever mode it was in when the power interruption occurred.

SETTING UP THE SYSTEM

As has been explained earlier in this chapter, the location of the day tank and storage tank will be governed by the position of the collector. If the latter is sited at ground level, the two tanks will, of necessity, be installed in a basement; if the collector is placed on the roof, the day and storage tanks may be at ground level or in a basement, whichever is most suitable.

The collector described has a surface area of 192 square feet, and will therefore serve a storage tank of approximately 400 gallons. The day tank recommended for this installation has a capacity of 65 gallons, and contains a heat-exchanger which may be connected to any desired domestic service.

Plumbing and Wiring

Plumbing connections between the Day Tank, the Storage Tank, and the Collector may be made in standard copper tube or P.V.C. If the latter is adopted, adapters will be necessary since the control valves on the Day Tank are designed for use with copper tube only.

NOTE: When sweating copper tube to the valves, silver solder must NOT be used; the high temperature necessary may damage the valve.

Plumbing connections required are as follows:

Valve #1 to collector input

Valve #3 to collector output. A "P" trap must be included in this line, located just below roof level inside the building. When the tank contains warm water (at a temperature higher than that of the collector, which has drained down) the trap will prevent migration of water vapour to the collector, where it would condense and might subsequently freeze.

Valve #2 and #4 via a common pipe to the Storage Tank inlet.

Storage Tank outlet to top connection on the Day Tank.

Heat exchanger connections to domestic service as required.

Good plumbing practice should be strictly observed, with no unnecessary elbows. Where the pipes are exposed above roof or ground level, they should be fully insulated with Armaflex (or equivalent) insulation. This insulation should be painted to prevent deterioration through exposure.

The Day Tank unit is delivered with the circuit board installed, and all necessary connections between the valves, the day tank temperature sensors and the board completed. A wiring harness, cut to the appropriate length as determined by the customer's requirement, provides connections with the collector, the storage tank, the day tank temperature control unit (T$_3$) and the power supply.

Complete instructions for installation, together with a wiring diagram, are included in the package. Although the electrical hook-up is relatively simple, local building codes in some areas may require that installation be performed by a licensed electrician, particularly if any extension of existing house wiring is required.

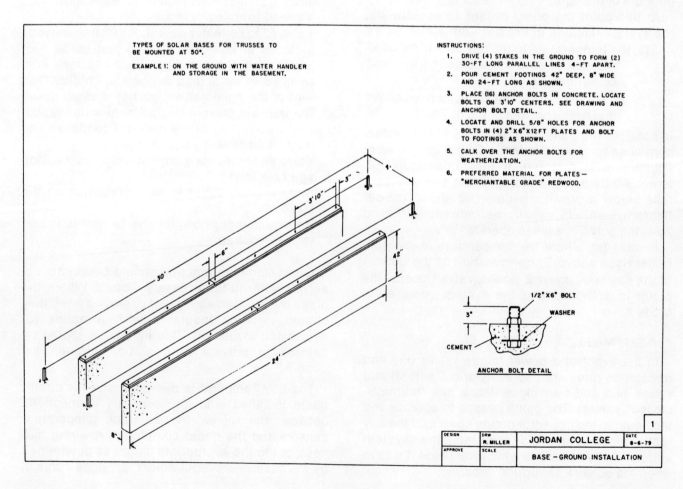

TYPES OF SOLAR BASES FOR TRUSSES TO BE MOUNTED AT 50°.

EXAMPLE 1: ON THE GROUND WITH WATER HANDLER AND STORAGE IN THE BASEMENT.

INSTRUCTIONS:

1. DRIVE (4) STAKES IN THE GROUND TO FORM (2) 30-FT LONG PARALLEL LINES 4-FT APART.
2. POUR CEMENT FOOTINGS 42" DEEP, 8" WIDE AND 24-FT LONG AS SHOWN.
3. PLACE (16) ANCHOR BOLTS IN CONCRETE. LOCATE BOLTS ON 3'10" CENTERS. SEE DRAWING AND ANCHOR BOLT DETAIL.
4. LOCATE AND DRILL 5/8" HOLES FOR ANCHOR BOLTS IN (4) 2"X6"X12FT PLATES AND BOLT TO FOOTINGS AS SHOWN.
5. CALK OVER THE ANCHOR BOLTS FOR WEATHERIZATION.
6. PREFERRED MATERIAL FOR PLATES — "MERCHANTABLE GRADE" REDWOOD.

ANCHOR BOLT DETAIL

JORDAN COLLEGE — BASE-GROUND INSTALLATION
DRW: R. MILLER DATE: 8-6-79

TYPES OF SOLAR BASES WITH TRUSSES TO BE MOUNTED ON ROOF.

EXAMPLE 2: ON A FLAT OR A PITCHED ROOF NOTE: FOR PITCHED ROOF DETERMINE THE ANGLE OF THE PITCH AND BUILD PLATFORM TO CARRY TRUSS ASSEMBLY AS SHOWN IN DRAWING NO. 4.

INSTRUCTIONS:

1. SNAP (2) PARALLEL CHALK LINES 28-FT LONG AND 4-FT APART.
2. CUT (14) 2"X4" SPACERS 12" LONG, LOCATE AND AND TACK TO ROOF ON 4-FT CENTERS.
3. TACK THE 2"X6" PLATES TO THE SPACERS.
4. DRILL 5/8" HOLES THROUGH PLATES, SPACERS AND ROOF AS SHOWN.
5. INSERT ANCHOR BOLTS IN HOLES, SLIP ROOF SUPPORTS IN PLACE AND TIGHTEN NUTS ON BOLTS. SEE ANCHOR BOLT DETAIL.
6. APPLY CALK TO EXTERIOR BOLTS OR NUTS.
7. PROTECT THE 2"X4" SPACERS BY APPLYING METAL FLASHING IN ACCORDANCE WITH PROPER BUILDING PROCEDURES.

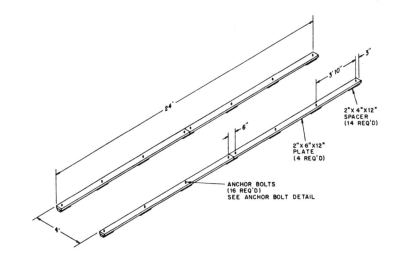

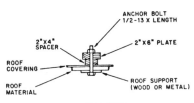

ANCHOR BOLT DETAIL

| DESIGN | DRW R MILLER | JORDAN COLLEGE | DATE 8-7-79 |
| APPROVE | SCALE | BASE - ROOF INSTALLATION | 2 |

INSTRUCTIONS:

1. THIRTEEN TRUSSES ARE NEEDED TO FRAME THE COLLECTOR.
2. EACH TRUSS MUST BE ON CENTER 24" FROM THE NEXT TRUSS.
3. METAL OR WOOD GUSSETS MUST BE NAILED SECURELY TO ALL THREE CORNERS OF THE TRUSS.
4. BUILD ONE TRUSS AND MEASURE TO INSURE CORRECT LENGTHS AND USE IT FOR A PATTERN TO BUILD (12) MORE TRUSSES FOR A TOTAL OF (13).
5. IF COLLECTOR BOX IS TO BE GLASS COVERED, BUILD TRUSSES IN ACCORDANCE WITH DETAIL "B".

BUILDING THE TRUSS AT A 50° ANGLE

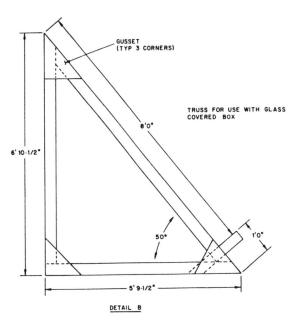

DETAIL A

DETAIL B

TRUSS FOR USE WITH GLASS COVERED BOX

NOTE: MAKE FROM REDWOOD-"MERCHANTABLE GRADE"

| DESIGN | DRW R MILLER | JORDAN COLLEGE | DATE 8-8-79 |
| APPROVE | SCALE | TRUSS 50° | 3 |

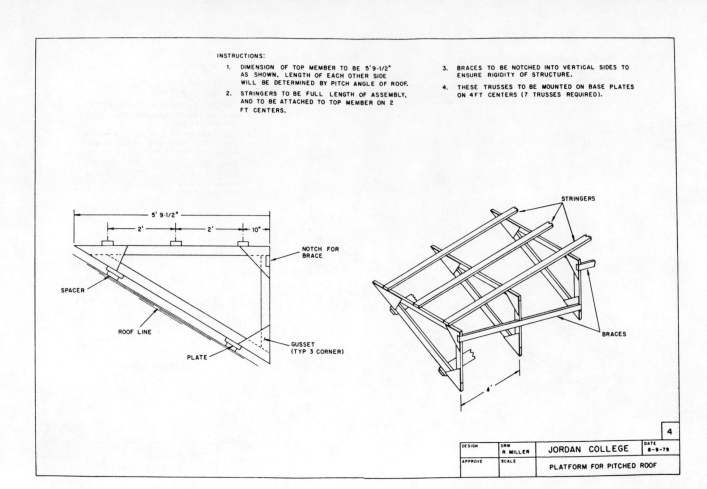

INSTRUCTIONS:

1. DIMENSION OF TOP MEMBER TO BE 5'9-1/2" AS SHOWN. LENGTH OF EACH OTHER SIDE WILL BE DETERMINED BY PITCH ANGLE OF ROOF.
2. STRINGERS TO BE FULL LENGTH OF ASSEMBLY, AND TO BE ATTACHED TO TOP MEMBER ON 2 FT CENTERS.
3. BRACES TO BE NOTCHED INTO VERTICAL SIDES TO ENSURE RIGIDITY OF STRUCTURE.
4. THESE TRUSSES TO BE MOUNTED ON BASE PLATES ON 4FT CENTERS (7 TRUSSES REQUIRED).

JORDAN COLLEGE — PLATFORM FOR PITCHED ROOF — DRW R. MILLER — DATE 8-9-79 — 4

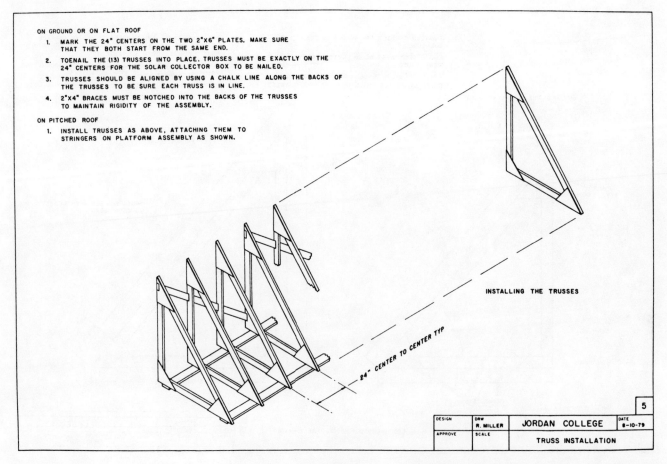

ON GROUND OR ON FLAT ROOF

1. MARK THE 24" CENTERS ON THE TWO 2"x6" PLATES. MAKE SURE THAT THEY BOTH START FROM THE SAME END.
2. TOENAIL THE (13) TRUSSES INTO PLACE. TRUSSES MUST BE EXACTLY ON THE 24" CENTERS FOR THE SOLAR COLLECTOR BOX TO BE NAILED.
3. TRUSSES SHOULD BE ALIGNED BY USING A CHALK LINE ALONG THE BACKS OF THE TRUSSES TO BE SURE EACH TRUSS IS IN LINE.
4. 2"x4" BRACES MUST BE NOTCHED INTO THE BACKS OF THE TRUSSES TO MAINTAIN RIGIDITY OF THE ASSEMBLY.

ON PITCHED ROOF

1. INSTALL TRUSSES AS ABOVE, ATTACHING THEM TO STRINGERS ON PLATFORM ASSEMBLY AS SHOWN.

INSTALLING THE TRUSSES

JORDAN COLLEGE — TRUSS INSTALLATION — DRW R. MILLER — DATE 8-10-79 — 5

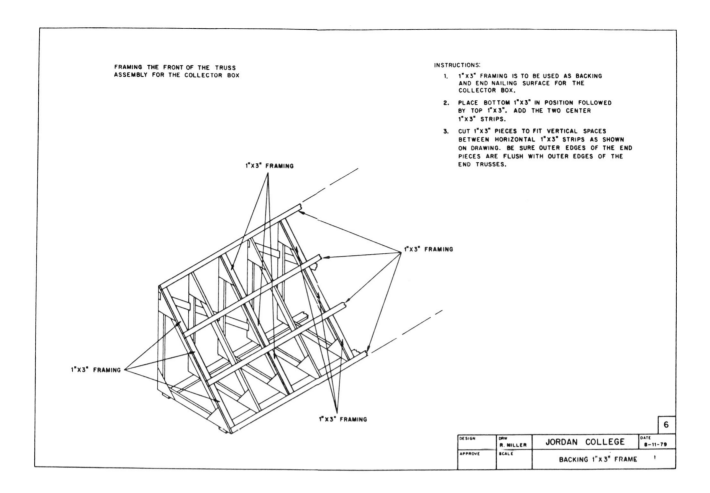

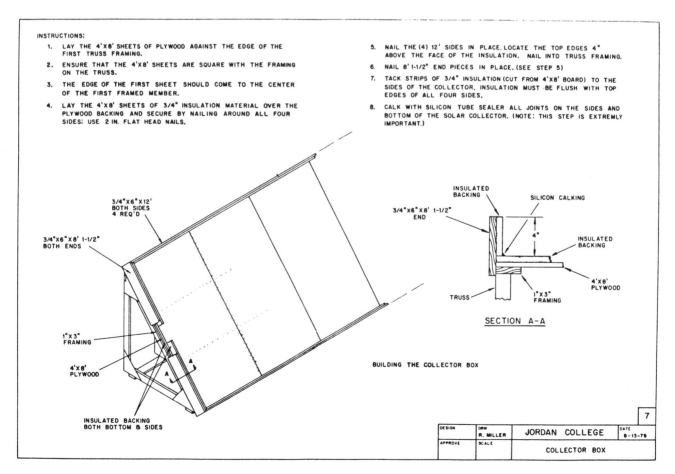

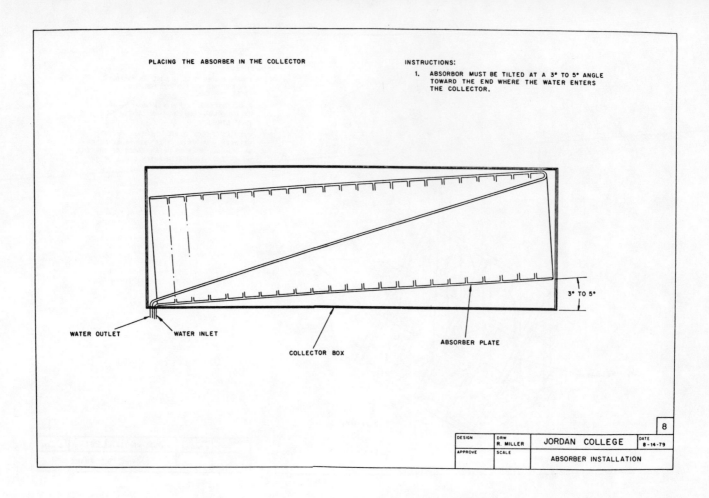

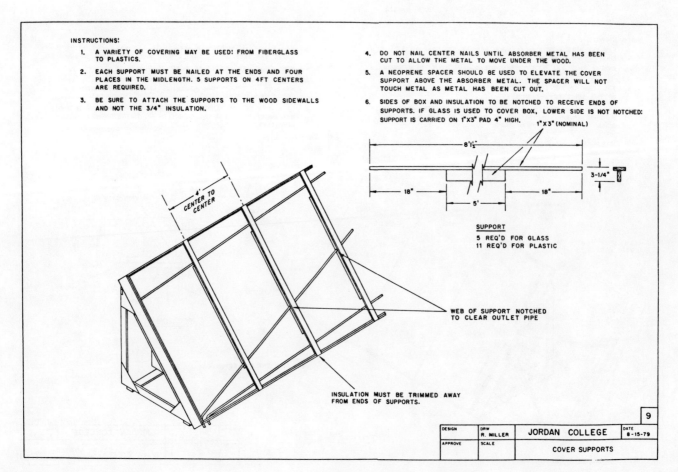

A

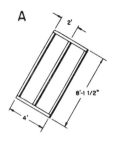

COLLECTOR COVER FRAME
MAKE (6) FROM 1"X2" STOCK.

B

APPLY SILICONE SEALANT AROUND
THE PERIMETER OF THE FRAME (DET A)
AND POSITION COVER MATERIAL
AS SHOWN. STAPLE AT 6" CENTERS.

C

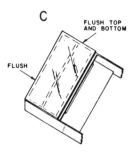

FLUSH TOP AND BOTTOM
FLUSH

APPLY SEALANT AROUND EDGE OF
COLLECTOR AND TOP OF BAFFLE-
COVER SUPPORT, POSITION DET B
COVER AS SHOWN AND FASTEN WITH
NO. 12 X 1 1/2" FLUSH HEAD WOOD SCREWS.
LOCATE SCREWS ON 12" CENTERS
ALONG TOP AND SIDES AND 6" CENTERS
ALONG BOTTOM. PREDRILL HOLES
FOR SCREWS. REPEAT FOR 6 COVERS.

D

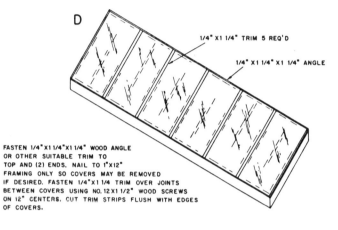

1/4" X 1 1/4" TRIM 5 REQ'D
1/4" X 1 1/4" X 1 1/4" ANGLE

FASTEN 1/4" X 1 1/4" X 1 1/4" WOOD ANGLE
OR OTHER SUITABLE TRIM TO
TOP AND (2) ENDS. NAIL TO 1"X12"
FRAMING ONLY SO COVERS MAY BE REMOVED
IF DESIRED. FASTEN 1/4" X 1 1/4 TRIM OVER JOINTS
BETWEEN COVERS USING NO. 12 X 1 1/2" WOOD SCREWS
ON 12" CENTERS. CUT TRIM STRIPS FLUSH WITH EDGES
OF COVERS.

DRW: R. MILLER
JORDAN COLLEGE
DATE: 8-21-79
COLLECTION BOX COVERS PLASTIC
10

INSTALLING THE GLASS

INSTRUCTIONS:

1. NAIL A 1"X3" STRIP ALONG THE UPPER SIDE AND BOTH ENDS OF THE COLLECTOR BOX AS SHOWN. NAIL TO THE SIDES OF THE COLLECTOR BOX BUT NOT TO THE COVER SUPPORTS.
2. NAIL A SECOND 1"X3" STRIP TO EACH COVER SUPPORT TO MAKE THEM FLUSH WITH OUTER 1"X3" BOARDS.
3. MEASURE GLASS AND SNAP A CHALK LINE ONTO THE OUTER EDGE OF COLLECTOR BOX TO INDICATE LOCATION OF GLASS. (ALLOW FOR GLASS TO BE SEATED 1/4" IN GROOVE IN LOWER SIDE.)
4. STRIP ON A FURRING STRIP 1/16" THICKER THAN THE GLASS AND LOCATE 1/4" OUTSIDE THE CHALK LINE. SEE DRAWING.
5. STRIP A FURRING STRIP ON EACH COVER SUPPORT. ALLOW FOR 1/4" SPACE BETWEEN FURRING STRIP AND GLASS COVERS ON BOTH SIDES OF THE STRIP.
6. LAY IN GLASS COVERS AND SEAL ALL EDGES WITH SILICONE SEALANT
7. USE APPROPRIATE WIDTH WOOD OR METAL STRIPS TO COVER THE JOINTS AND FASTEN IN PLACE WITH SCREWS. SEE DRAWING.
8. INSULATION APPLIED TO LOWER SIDE TO BE 4-3/4" WIDE (FLUSH WITH INNER EDGE OF GLASS GROOVE).

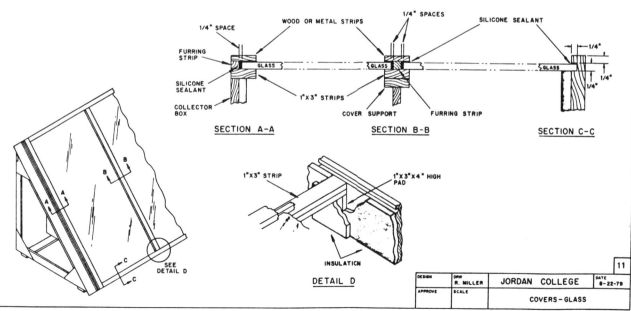

DRW: R. MILLER
JORDAN COLLEGE
DATE: 8-22-79
COVERS - GLASS
11

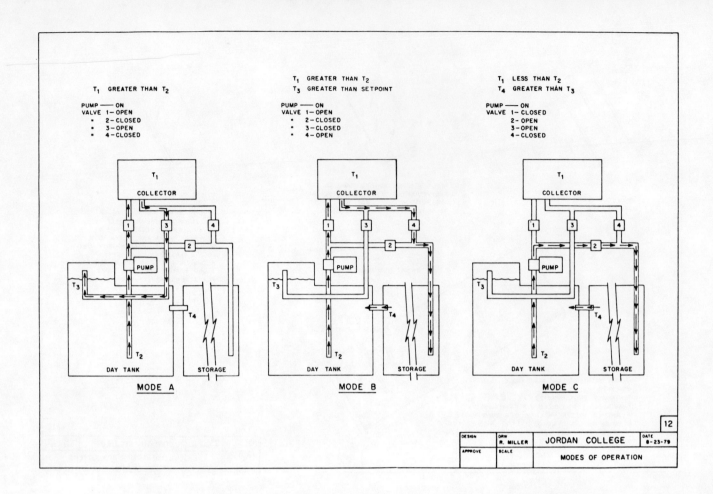

Appendix A:

Weather Data for the United States

Average Monthly and Yearly Degree-Days for the United States

Solar Politions and Insolation Values for Various Latitudes

Design Values of Various Building and Insulation Materials

Coefficients of Transmission (U) for Various Structural Elements

Weather Design Data for the United States [a,b,c]

State and Station	Lat. °	Winter Median of Annual Extremes	99%	97½%	Coincident Wind Velocity	State and Station	Lat. °	Winter Median of Annual Extremes	99%	97½%	Coincident Wind Velocity
ALABAMA						**CALIFORNIA** (continued)					
Alexander City	33	12	16	20	L	Livermore	37	23	28	30	VL
Anniston AP	33	12	17	19	L	Lompoc, Vandenburg AFB	34	32	36	38	VL
Auburn	32	17	21	25	L	Long Beach AP	33	31	36	38	VL
Birmingham AP	33	14	19	22	L	Los Angeles AP	34	36	41	43	VL
Decatur	34	10	15	19	L	Los Angeles CO	34	38	42	44	VL
Dothan AP	31	19	23	27	L	Merced-Castle AFB	37	24	30	32	VL
Florence AP	34	8	13	17	L	Modesto	37	26	32	36	VL
Gadsden	34	11	16	20	L	Monterey	36	29	34	37	VL
Huntsville AP	34	8	13	17	L	Napa	38	26	31	34	VL
Mobile AP	30	21	26	29	M	Needles AP	34	27	33	37	VL
Mobile CO	30	24	28	32	M	Oakland AP	37	30	35	37	VL
Montgomery AP	32	18	22	26	L	Oceanside	33	33	38	40	VL
Selma-Craig AFB	32	18	23	27	L	Ontario	34	26	32	34	VL
Talladega	33	11	15	19	L	Oxnard AFB	34	32	35	37	VL
Tuscaloosa AP	33	14	19	23	L	Palmdale AP	34	18	24	27	VL
ALASKA						Palm Springs	33	27	32	36	VL
Anchorage AP	61	−29	−25	−20	VL	Pasadena	34	31	36	39	VL
Barrow	71	−49	−45	−42	M	Petaluma	38	24	29	32	VL
Fairbanks AP	64	−59	−53	−50	VL	Pomona CO	34	26	31	34	VL
Juneau AP	58	−11	−7	−4	L	Redding AP	40	25	31	35	VL
Kodiak	57	4	8	12	M	Redlands	34	28	34	37	VL
Nome AP	64	−37	−32	−28	L	Richmond	38	28	35	38	VL
ARIZONA†						Riverside-March AFB	33	26	32	34	VL
Douglas AP	31	13	18	22	VL	Sacramento AP	38	24	30	32	VL
Flagstaff AP	35	−10	0	5	VL	Salinas AP	36	27	32	35	VL
Fort Huachuca AP	31	18	25	28	VL	San Bernardino, Norton AFB	34	26	31	33	VL
Kingman AP	35	18	25	29	VL	San Diego AP	32	38	42	44	VL
Nogales	31	15	20	24	VL	San Fernando	34	29	34	37	VL
Phoenix AP	33	25	31	34	VL	San Francisco AP	37	32	35	37	L
Prescott AP	34	7	15	19	VL	San Francisco CO	37	38	42	44	VL
Tucson AP	33	23	29	32	VL	San Jose AP	37	30	34	36	VL
Winslow AP	35	2	9	13	VL	San Luis Obispo	35	30	35	37	VL
Yuma AP	32	32	37	40	VL	Santa Ana AP	33	28	33	36	VL
ARKANSAS						Santa Barbara CO	34	30	34	36	VL
Blytheville AFB	36	6	12	17	L	Santa Cruz	37	28	32	34	VL
Camden	33	13	19	23	L	Santa Maria AP	34	28	32	34	VL
El Dorado AP	33	13	19	23	L	Santa Monica CO	34	38	43	45	VL
Fayetteville AP	36	3	9	13	M	Santa Paula	34	28	33	36	VL
Fort Smith AP	35	9	15	19	M	Santa Rosa	38	24	29	32	VL
Hot Springs Nat. Pk.	34	12	18	22	M	Stockton AP	37	25	30	34	VL
Jonesboro	35	8	14	18	M	Ukiah	39	22	27	30	VL
Little Rock AP	34	13	19	23	M	Visalia	36	26	32	36	VL
Pine Bluff AP	34	14	20	24	L	Yreka	41	7	13	17	VL
Texarkana AP	33	16	22	26	M	Yuba City	39	24	30	34	VL
CALIFORNIA						**COLORADO**					
Bakersfield AP	35	26	31	33	VL	Alamosa AP	37	−26	−17	−13	VL
Barstow AP	34	18	24	28	VL	Boulder	40	−5	4	8	L
Blythe AP	33	26	31	35	VL	Colorado Springs AP	38	−9	−1	4	L
Burbank AP	34	30	36	38	VL	Denver AP	39	−9	−2	3	L
Chico	39	23	29	33	VL	Durango	37	−10	0	4	VL
Concord	38	27	32	36	VL	Fort Collins	40	−16	−9	−5	L
Covina	34	32	38	41	VL	Grand Junction AP	39	−2	8	11	VL
Crescent City AP	41	28	33	36	L	Greeley	40	−18	−9	−5	L
Downey	34	30	35	38	VL	La Junta AP	38	−14	−6	−2	M
El Cajon	32	26	31	34	VL	Leadville	39	−18	−9	−4	VL
El Centro AP	32	26	31	35	VL	Pueblo AP	38	−14	−5	−1	L
Escondido	33	28	33	36	VL	Sterling	40	−15	−6	−2	M
Eureka/Arcata AP	41	27	32	35	L	Trinidad AP	37	−9	1	5	L
Fairfield-Travis AFB	38	26	32	34	VL						
Fresno AP	36	25	28	31	VL						
Hamilton AFB	38	28	33	35	VL						
Laguna Beach	33	32	37	39	VL						

Notes:
a. From ASHRAE *Handbook of Fundamentals*.
b. AP indicates airport; AFB indicates Air Force base; CO indicates cosmopolitan area; other may be taken to be semi-rural.
c. Wind velocity: VL = Very Light, 70% or more of cold extreme hours at less than 7 mph; L = Light, 50-69% cold extreme hours less than 7 mph; M = Moderate, 50-74% cold extreme hours more than 7 mph; H = High, 75% or more cold extreme hours more than 7 mph, 50% more than 12 mph.

State and Station	Lat. °	Median of Annual Extremes	Winter 99%	Winter 97½%	Coincident Wind Velocity
CONNECTICUT					
Bridgeport AP	41	−1	4	8	M
Hartford, Brainard Field	41	−4	1	5	M
New Haven AP	41	0	5	9	H
New London	41	0	4	8	H
Norwalk	41	−5	0	4	M
Norwich	41	−7	−2	2	M
Waterbury	41	−5	0	4	M
Windsor Locks, Bradley Field	42	−7	−2	2	M
DELAWARE					
Dover AFB	39	8	13	15	M
Wilmington AP	39	6	12	15	M
DISTRICT OF COLUMBIA					
Andrews AFB	38	9	13	16	M
Washington National AP	38	12	16	19	M
FLORIDA					
Belle Glade	26	31	35	39	M
Cape Kennedy AP	28	33	37	40	L
Daytona Beach AP	29	28	32	36	L
Fort Lauderdale	26	37	41	45	M
Fort Myers AP	26	34	38	42	M
Fort Pierce	27	33	37	41	M
Gainesville AP	29	24	28	32	L
Jacksonville AP	30	26	29	32	L
Key West AP	24	50	55	58	M
Lakeland CO	28	31	35	39	M
Miami AP	25	39	44	47	M
Miami Beach CO	25	40	45	48	M
Ocala	29	25	29	33	L
Orlando AP	28	29	33	37	L
Panama City, Tyndall AFB	30	28	32	35	M
Pensacola CO	30	25	29	32	M
St. Augustine	29	27	31	35	L
St. Petersburg	28	35	39	42	M
Sanford	28	29	33	37	L
Sarasota	27	31	35	39	M
Tallahassee AP	30	21	25	29	L
Tampa AP	28	32	36	39	M
West Palm Beach AP	26	36	40	44	M
GEORGIA					
Albany, Turner AFB	31	21	26	30	L
Americus	32	18	22	25	L
Athens	34	12	17	21	L
Atlanta AP	33	14	18	23	H
Augusta AP	33	17	20	23	L
Brunswick	31	24	27	31	L
Columbus, Lawson AFB	32	19	23	26	L
Dalton	34	10	15	19	L
Dublin	32	17	21	25	L
Gainesville	34	11	16	20	L
Griffin	33	13	17	22	L
La Grange	33	12	16	20	L
Macon AP	32	18	23	27	L
Marietta, Dobbins AFB	34	12	17	21	L
Moultrie	31	22	26	30	L
Rome AP	34	11	16	20	L
Savannah-Travis AP	32	21	24	27	L
Valdosta-Moody AFB	31	24	28	31	L
Waycross	31	20	24	28	L
HAWAII					
Hilo AP	19	56	59	61	L
Honolulu AP	21	58	60	62	L
Kaneohe	21	58	60	61	L
Wahiawa	21	57	59	61	L
IDAHO					
Boise AP	43	0	4	10	L
Burley	42	−5	4	8	VL
Coeur d'Alene AP	47	−4	2	7	VL
Idaho Falls AP	43	−17	−12	−6	VL
Lewiston AP	46	1	6	12	VL
Moscow	46	−11	−3	1	VL
Mountain Home AFB	43	−3	2	9	L
Pocatello AP	43	−12	−8	−2	VL
Twin Falls AP	42	−5	4	8	L
ILLINOIS					
Aurora	41	−13	−7	−3	M
Belleville, Scott AFB	38	0	6	10	M
Bloomington	40	−7	−1	3	M
Carbondale	37	1	7	11	M
Champaign/Urbana	40	−6	0	4	M
Chicago, Midway AP	41	−7	−4	1	M
Chicago, O'Hare AP	42	−9	−4	0	M
Chicago, CO	41	−5	−3	1	M
Danville	40	−6	−1	4	M
Decatur	39	−6	0	4	M
Dixon	41	−13	−7	−3	M
Elgin	42	−14	−8	−4	M
Freeport	42	−16	−10	−6	M
Galesburg	41	−10	−4	0	M
Greenville	39	−3	3	7	M
Joliet AP	41	−11	−5	−1	M
Kankakee	41	−10	−4	1	M
La Salle/Peru	41	−9	−3	1	M
Macomb	40	−5	−3	1	M
Moline AP	41	−12	−7	−3	M
Mt. Vernon	38	0	6	10	M
Peoria AP	40	−8	−2	2	M
Quincy AP	40	−8	−2	2	M
Rantoul, Chanute AFB	40	−7	−1	3	M
Rockford	42	−13	−7	−3	M
Springfield AP	39	−7	−1	4	M
Waukegan	42	−11	−5	−1	M
INDIANA					
Anderson	40	−5	0	5	M
Bedford	38	−3	3	7	M
Bloomington	39	−3	3	7	M
Columbus, Bakalar AFB	39	−3	3	7	M
Crawfordsville	40	−8	−2	2	M
Evansville AP	38	1	6	10	M
Fort Wayne AP	41	−5	0	5	M
Goshen AP	41	−10	−4	0	M
Hobart	41	−10	−4	0	M
Huntington	40	−8	−2	2	M
Indianapolis AP	39	−5	0	4	M
Jeffersonville	38	3	9	13	M
Kokomo	40	−6	0	4	M
Lafayette	40	−7	−1	3	M
La Porte	41	−10	−4	0	M
Marion	40	−8	−2	2	M
Muncie	40	−8	−2	2	M
Peru, Bunker Hill AFB	40	−9	−3	1	M
Richmond AP	39	−7	−1	3	M
Shelbyville	39	−4	2	6	M
South Bend AP	41	−6	−2	3	M
Terre Haute AP	39	−3	3	7	M
Valparaiso	41	−12	−6	−2	M
Vincennes	38	−1	5	9	M

State and Station	Lat. °	Winter Median of Annual Extremes	99%	97½%	Coincident Wind Velocity	State and Station	Lat. °	Winter Median of Annual Extremes	99%	97½%	Coincident Wind Velocity
IOWA						**MARYLAND**					
Ames	42	−17	−11	−7	M	Baltimore AP	39	8	12	15	M
Burlington AP	40	−10	−4	0	M	Baltimore CO	39	12	16	20	M
Cedar Rapids AP	41	−14	−8	−4	M	Cumberland	39	0	5	9	L
Clinton	41	−13	−7	−3	M	Frederick AP	39	2	7	11	M
Council Bluffs	41	−14	−7	−3	M	Hagerstown	39	1	6	10	L
Des Moines AP	41	−13	−7	−3	M	Salisbury	38	10	14	18	M
Dubuque	42	−17	−11	−7	M	**MASSACHUSETTS**					
Fort Dodge	42	−18	−12	−8	M	Boston AP	42	−1	6	10	H
						Clinton	42	−8	−2	2	M
Iowa City	41	−14	−8	−4	M	Fall River	41	−1	5	9	H
Keokuk	40	−9	−3	1	M	Framingham	42	−7	−1	3	M
Marshalltown	42	−16	−10	−6	M	Gloucester	42	−4	2	6	H
Mason City AP	43	−20	−13	−9	M						
						Greenfield	42	−12	−6	−2	M
Newton	41	−15	−9	−5	M	Lawrence	42	−9	−3	1	M
Ottumwa AP	41	−12	−6	−2	M	Lowell	42	−7	−1	3	M
Sioux City AP	42	−17	−10	−6	M	New Bedford	41	3	9	13	H
Waterloo	42	−18	−12	−8	M						
KANSAS						Pittsfield AP	42	−11	−5	−1	M
Atchison	39	−9	−2	2	M	Springfield, Westover AFB	42	−8	−3	2	M
Chanute AP	37	−3	3	7	H	Taunton	41	−9	−4	0	H
Dodge City AP	37	−5	3	7	M	Worcester AP	42	−8	−3	1	M
El Dorado	37	−3	4	8	H	**MICHIGAN**					
Emporia	38	−4	3	7	H	Adrian	41	−6	0	4	M
						Alpena AP	45	−11	−5	−1	M
Garden City AP	38	−10	−1	3	M	Battle Creek AP	42	−6	1	5	M
Goodland AP	39	−10	−2	4	M	Benton Harbor AP	42	−7	−1	3	M
Great Bend	38	−5	2	6	M	Detroit Met. CAP	42	0	4	8	M
Hutchinson AP	38	−5	2	6	M	Escanaba	45	−13	−7	−3	M
Liberal	37	−4	4	8	M						
						Flint AP	43	−7	−1	3	M
Manhattan, Fort Riley	39	−7	−1	4	H	Grand Rapids AP	42	−2	2	6	M
Parsons	37	−2	5	9	H	Holland	42	−4	2	6	M
Russell AP	38	−7	0	4	M	Jackson AP	42	−6	0	4	M
Salina	38	−4	3	7	H	Kalamazoo	42	−5	1	5	M
Topeka AP	39	−4	3	6	M						
Wichita AP	37	−1	5	9	H	Lansing AP	42	−4	2	6	M
KENTUCKY						Marquette CO	46	−14	−8	−4	L
Ashland	38	1	6	10	L	Mt. Pleasant	43	−9	−3	1	M
Bowling Green AP	37	1	7	11	L	Muskegon AP	43	−2	4	8	M
Corbin AP	37	0	5	9	L	Pontiac	42	−6	0	4	M
Covington AP	39	−3	3	8	L						
Hopkinsville, Campbell AFB	36	4	10	14	L	Port Huron	43	−6	−1	3	M
						Saginaw AP	43	−7	−1	3	M
Lexington AP	38	0	6	10	M	Sault Ste. Marie AP	46	−18	−12	−8	L
Louisville AP	38	1	8	12	L	Traverse City AP	44	−6	0	4	M
Madisonville	37	1	7	11	L	Ypsilanti	42	−3	−1	5	M
Owensboro	37	0	6	10	L	**MINNESOTA**					
Paducah AP	37	4	10	14	L	Albert Lea	43	−20	−14	−10	M
LOUISIANA						Alexandria AP	45	−26	−19	−15	L
Alexandria AP	31	20	25	29	L	Bemidji AP	47	−38	−32	−28	L
Baton Rouge AP	30	22	25	30	L	Brainerd	46	−31	−24	−20	L
Bogalusa	30	20	24	28	L						
Houma	29	25	29	33	L	Duluth AP	46	−25	−19	−15	M
						Faribault	44	−23	−16	−12	L
Lafayette AP	30	23	28	32	L	Fergus Falls	46	−28	−21	−17	L
Lake Charles AP	30	25	29	33	M	International Falls AP	48	−35	−29	−24	L
Minden	32	17	22	26	L						
Monroe AP	32	18	23	27	L	Mankato	44	−23	−16	−12	L
Natchitoches	31	17	22	26	L	Minneapolis/St. Paul AP	44	−19	−14	−10	M
New Orleans AP	30	29	32	35	M	Rochester AP	44	−23	−17	−13	M
Shreveport AP	32	18	22	26	M	St. Cloud AP	45	−26	−20	−16	L
MAINE						Virginia	47	−32	−25	−21	L
Augusta AP	44	−13	−7	−3	M	Willmar	45	−25	−18	−14	L
Bangor, Dow AFB	44	−14	−8	−4	M	Winona	44	−19	−12	−8	M
Caribou AP	46	−24	−18	−14	L	**MISSISSIPPI**					
Lewiston	44	−14	−8	−4	M	Biloxi, Keesler AFB	30	26	30	32	M
Millinocket AP	45	−22	−16	−12	L	Clarksdale	34	14	20	24	L
Portland AP	43	−14	−5	0	L	Columbus AFB	33	13	18	22	L
Waterville	44	−15	−9	−5	M	Greenville AFB	33	16	21	24	L

State and Station	Lat. °	Winter Median of Annual Extremes	99%	97½%	Coincident Wind Velocity	State and Station	Lat. °	Winter Median of Annual Extremes	99%	97½%	Coincident Wind Velocity
MISSISSIPPI (continued)						**NEVADA** (continued)					
Greenwood	33	14	19	23	L	Reno AP	39	−2	2	7	VL
Hattiesburg	31	18	22	26	L	Reno CO	39	8	12	17	VL
Jackson AP	32	17	21	24	L	Tonopah AP	38	2	9	13	VL
						Winnemucca AP	40	−8	1	5	VL
Laurel	31	18	22	26	L	**NEW HAMPSHIRE**					
McComb AP	31	18	22	26	L	Berlin	44	−25	−19	−15	L
Meridian AP	32	15	20	24	L	Claremont	43	−19	−13	−9	L
Natchez	31	18	22	26	L	Concord AP	43	−17	−11	−7	M
Tupelo	34	13	18	22	L	Keene	43	−17	−12	−8	M
Vicksburg CO	32	18	23	26	L	Laconia	43	−22	−16	−12	M
MISSOURI						Manchester, Grenier AFB	43	−11	−5	1	M
Cape Girardeau	37	2	8	12	M	Portsmouth, Pease AFB	43	−8	−2	3	M
Columbia AP	39	−4	2	6	M	**NEW JERSEY**					
Farmington AP	37	−2	4	8	M	Atlantic City CO	39	10	14	18	H
Hannibal	39	−7	−1	4	M	Long Branch	40	4	9	13	H
Jefferson City	38	−4	2	6	M	Newark AP	40	6	11	15	M
Joplin AP	37	1	7	11	M	New Brunswick	40	3	8	12	M
Kansas City AP	39	−2	4	8	M	Paterson	40	3	8	12	M
						Phillipsburg	40	1	6	10	L
Kirksville AP	40	−13	−7	−3	M	Trenton CO	40	7	12	16	M
Mexico	39	−7	−1	3	M	Vineland	39	7	12	16	M
Moberly	39	−8	−2	2	M	**NEW MEXICO**					
Poplar Bluff	36	3	9	13	M	Alamagordo, Holloman AFB	32	12	18	22	L
Rolla	38	−3	3	7	M	Albuquerque AP	35	6	14	17	L
						Artesia	32	9	16	19	L
						Carlsbad AP	32	11	17	21	L
St. Joseph AP	39	−8	−1	3	M	Clovis AP	34	2	14	17	L
St. Louis AP	38	−2	4	8	M	Farmington AP	36	−3	6	9	VL
St. Louis CO	38	1	7	11	M						
Sedalia, Whiteman AFB	38	−2	4	9	M	Gallup	35	−13	−5	−1	VL
Sikeston	36	4	10	14	L	Grants	35	−15	−7	−3	VL
Springfield AP	37	0	5	10	M	Hobbs AP	32	9	15	19	L
MONTANA						Las Cruces	32	13	19	23	L
Billings AP	45	−19	−10	−6	L	Los Alamos	35	−4	5	9	L
Bozeman	45	−25	−15	−11	L	Raton AP	36	−11	−2	2	L
Butte AP	46	−34	−24	−16	VL						
Cut Bank AP	48	−32	−23	−17	L	Roswell, Walker AFB	33	5	16	19	L
Glasgow AP	48	−33	−25	−20	L	Santa Fe CO	35	−2	7	11	L
Glendive	47	−28	−20	−16	L	Silver City AP	32	8	14	18	VL
Great Falls AP	47	−29	−20	−16	L	Socorro AP	34	6	13	17	L
						Tucumcari AP	35	1	9	13	L
Havre	48	−32	−22	−15	M	**NEW YORK**					
Helena AP	46	−27	−17	−13	L	Albany AP	42	−14	−5	0	L
Kalispell AP	48	−17	−7	−3	VL	Albany CO	42	−5	1	5	L
Lewiston AP	47	−27	−18	−14	L	Auburn	43	−10	−2	2	M
Livingston AP	45	−26	−17	−13	L	Batavia	43	−7	−1	3	M
Miles City AP	46	−27	−19	−15	L	Binghamton CO	42	−8	−2	2	L
Missoula AP	46	−16	−7	−3	VL						
NEBRASKA						Buffalo AP	43	−3	3	6	M
Beatrice	40	−10	−3	1	M	Cortland	42	−11	−5	−1	L
Chadron AP	42	−21	−13	−9	M	Dunkirk	42	−2	4	8	M
Columbus	41	−14	−7	−3	M	Elmira AP	42	−5	1	5	L
Fremont	41	−14	−7	−3	M	Geneva	42	−8	−2	2	M
Grand Island AP	41	−14	−6	−2	M	Glens Falls	43	−17	−11	−7	L
Hastings	40	−11	−3	1	M	Gloversville	43	−12	−6	−2	L
Kearney	40	−14	−6	−2	M	Hornell	42	−15	−9	−5	L
Lincoln CO	40	−10	−4	0	M						
McCook	40	−12	−4	0	M	Ithaca	42	−10	−4	0	L
Norfolk	42	−18	−11	−7	M	Jamestown	42	−5	1	5	M
North Platte AP	41	−13	−6	−2	M	Kingston	42	−8	−2	2	L
Omaha AP	41	−12	−5	−1	M	Lockport	43	−4	2	6	M
Scottsbluff AP	41	−16	−8	−4	M	Massena AP	45	−22	−16	−12	M
Sidney AP	41	−15	−7	−2	M						
NEVADA						Newburgh-Stewart AFB	41	−4	2	6	M
Carson City	39	−4	3	7	VL	NYC-Central Park	40	6	11	15	H
Elko AP	40	−21	−13	−7	VL	NYC-Kennedy AP	40	12	17	21	H
Ely AP	39	−15	−6	−2	VL	NYC-LaGuardia AP	40	7	12	16	H
Las Vegas AP	36	18	23	26	VL	Niagara Falls AP	43	−2	4	7	M
Lovelock AP	40	0	7	11	VL	Olean	42	−13	−8	−3	L

State and Station	Lat. °	Winter Median of Annual Extremes	99%	97½%	Coincident Wind Velocity	State and Station	Lat. °	Winter Median of Annual Extremes	99%	97½%	Coincident Wind Velocity
NEW YORK (continued)						**OHIO** (continued)					
Oneonta	42	−13	−7	−3	L	Steubenville	40	−2	4	9	M
Oswego CO	43	−4	2	6	M	Toledo AP	41	−5	1	5	M
Plattsburg AFB	44	−16	−10	−6	L	Warren	41	−6	0	4	M
Poughkeepsie	41	−6	−1	3	L	Wooster	40	−7	−1	3	M
Rochester AP	43	−5	2	5	M	Youngstown AP	41	−5	1	6	M
Rome-Griffiss AFB	43	−13	−7	−3	L	Zanesville AP	40	−7	−1	3	M
						OKLAHOMA					
Schenectady	42	−11	−5	−1	L	Ada	34	6	12	16	H
Suffolk County AFB	40	4	9	13	H	Altus AFB	34	7	14	18	H
Syracuse AP	43	−10	−2	2	M	Ardmore	34	9	15	19	H
Utica	43	−12	−6	−2	L	Bartlesville	36	−1	5	9	H
Watertown	44	−20	−14	−10	M	Chickasha	35	5	12	16	H
NORTH CAROLINA											
Asheville AP	35	8	13	17	L	Enid-Vance AFB	36	3	10	14	H
Charlotte AP	35	13	18	22	L	Lawton AP	34	6	13	16	H
Durham	36	11	15	19	L	McAlester	34	7	13	17	H
Elizabeth City AP	36	14	18	22	M	Muskogee AP	35	6	12	16	M
Fayetteville, Pope AFB	35	13	17	20	L	Norman	35	5	11	15	H
						Oklahoma City AP	35	4	11	15	H
Goldsboro, Seymour AFB	35	14	18	21	M						
Greensboro AP	36	9	14	17	L	Ponca City	36	1	8	12	H
Greenville	35	14	18	22	M	Seminole	35	6	12	16	H
Henderson	36	8	12	16	L	Stillwater	36	2	9	13	H
Hickory	35	9	14	18	L	Tulsa AP	36	4	12	16	H
Jacksonville	34	17	21	25	M	Woodward	36	−3	4	8	H
						OREGON					
Lumberton	34	14	18	22	L	Albany	44	17	23	27	VL
New Bern AP	35	14	18	22	L	Astoria AP	46	22	27	30	M
Raleigh/Durham AP	35	13	16	20	L	Baker AP	44	−10	−3	1	VL
Rocky Mount	36	12	16	20	L	Bend	44	−7	0	4	VL
Wilmington AP	34	19	23	27	L	Corvallis	44	17	23	27	VL
Winston-Salem AP	36	9	14	17	L						
NORTH DAKOTA						Eugene AP	44	16	22	26	VL
Bismarck AP	46	−31	−24	−19	VL	Grants Pass	42	16	22	26	VL
Devil's Lake	48	−30	−23	−19	M	Klamath Falls AP	42	−5	1	5	VL
Dickinson AP	46	−31	−23	−19	L	Medford AP	42	15	21	23	VL
Fargo AP	46	−28	−22	−17	L	Pendleton AP	45	−2	3	10	VL
Grand Forks AP	48	−30	−26	−23	L	Portland AP	45	17	21	24	L
Jamestown AP	47	−29	−22	−18	L	Portland CO	45	21	26	29	L
Minot AP	48	−31	−24	−20	M	Roseburg AP	43	19	25	29	VL
Williston	48	−28	−21	−17	M	Salem AP	45	15	21	25	VL
OHIO						The Dalles	45	7	13	17	VL
Akron/Canton AP	41	−5	1	6	M	**PENNSYLVANIA**					
Ashtabula	42	−3	3	7	M	Allentown AP	40	−2	3	5	M
Athens	39	−3	3	7	M	Altoona CO	40	−4	1	5	L
Bowling Green	41	−7	−1	3	M	Butler	40	−8	−2	2	L
Cambridge	40	−6	0	4	M	Chambersburg	40	0	5	9	L
						Erie AP	42	1	7	11	M
Chillicothe	39	−1	5	9	M						
Cincinnati CO	39	−2	8	12	L	Harrisburg AP	40	4	9	13	L
Cleveland AP	41	−2	2	7	M	Johnstown	40	−4	1	5	L
Columbus AP	40	−1	2	7	M	Lancaster	40	−3	2	6	L
Dayton AP	39	−2	0	6	M	Meadville	41	−6	0	4	M
						New Castle	41	−7	−1	4	M
Defiance	41	−7	−1	1	M	Philadelphia AP	39	7	11	15	M
Findlay AP	41	−6	0	4	M	Pittsburgh AP	40	−1	5	9	M
Fremont	41	−7	−1	3	M	Pittsburgh CO	40	1	7	11	M
Hamilton	39	−2	4	8	M	Reading CO	40	1	6	9	M
Lancaster	39	−5	1	5	M	Scranton/Wilkes-Barre	41	−3	2	6	L
Lima	40	−6	0	4	M	State College	40	−3	2	6	L
Mansfield AP	40	−7	1	3	M	Sunbury	40	−2	3	7	L
Marion	40	−5	1	6	M	Uniontown	39	−1	4	8	L
Middletown	39	−3	3	7	M	Warren	41	−8	−3	1	L
Newark	40	−7	−1	3	M	West Chester	40	4	9	13	M
						Williamsport AP	41	−5	1	5	L
Norwalk	41	−7	−1	3	M	York	40	−1	4	8	L
Portsmouth	38	0	5	9	L	**RHODE ISLAND**					
Sandusky CO	41	−2	4	8	M	Newport	41	1	5	11	H
Springfield	40	−3	3	7	M	Providence AP	41	0	6	10	M

64

State and Station	Lat. °	Median of Annual Extremes	Winter 99%	Winter 97½%	Coincident Wind Velocity
SOUTH CAROLINA					
Anderson	34	13	18	22	L
Charleston AFB	32	19	23	27	L
Charleston CO	32	23	26	30	L
Columbia AP	34	16	20	23	L
Florence AP	34	16	21	25	L
Georgetown	33	19	23	26	L
Greenville AP	34	14	19	23	L
Greenwood	34	15	19	23	L
Orangeburg	33	17	21	25	L
Rock Hill	35	13	17	21	L
Spartanburg AP	35	13	18	22	L
Sumter-Shaw AFB	34	18	23	26	L
SOUTH DAKOTA					
Aberdeen AP	45	−29	−22	−18	L
Brookings	44	−26	−19	−15	M
Huron AP	44	−24	−16	−12	L
Mitchell	43	−22	−15	−11	M
Pierre AP	44	−21	−13	−9	M
Rapid City AP	44	−17	−9	−6	M
Sioux Falls AP	43	−21	−14	−10	M
Watertown AP	45	−27	−20	−16	L
Yankton	43	−18	−11	−7	M
TENNESSEE					
Athens	33	10	14	18	L
Bristol-Tri City AP	36	6	11	16	L
Chattanooga AP	35	11	15	19	L
Clarksville	36	6	12	16	L
Columbia	35	8	13	17	L
Dyersburg	36	7	13	17	L
Greenville	35	5	10	14	L
Jackson AP	35	8	14	17	L
Knoxville AP	35	9	13	17	L
Memphis AP	35	11	17	21	L
Murfreesboro	35	7	13	17	L
Nashville AP	36	6	12	16	L
Tullahoma	35	7	13	17	L
TEXAS					
Abilene AP	32	12	17	21	M
Alice AP	27	26	30	34	M
Amarillo AP	35	2	8	12	M
Austin AP	30	19	25	29	M
Bay City	29	25	29	33	M
Beaumont	30	25	29	33	M
Beeville	28	24	28	32	M
Big Spring AP	32	12	18	22	M
Brownsville AP	25	32	36	40	M
Brownwood	31	15	20	25	M
Bryan AP	30	22	27	31	M
Corpus Christi AP	27	28	32	36	M
Corsicana	32	16	21	25	M
Dallas AP	32	14	19	24	H
Del Rio, Laughlin AFB	29	24	28	31	M
Denton	33	12	18	22	H
Eagle Pass	28	23	27	31	L
El Paso AP	31	16	21	25	L
Fort Worth AP	32	14	20	24	H
Galveston AP	29	28	32	36	M
Greenville	33	13	19	24	H
Harlingen	26	30	34	38	M
Houston AP	29	23	28	32	M
Houston CO	29	24	29	33	M
Huntsville	30	22	27	31	M

State and Station	Lat. °	Median of Annual Extremes	Winter 99%	Winter 97½%	Coincident Wind Velocity
TEXAS (continued)					
Killeen-Gray AFB	31	17	22	26	M
Lamesa	32	7	14	18	M
Laredo AFB	27	29	32	36	L
Longview	32	16	21	25	M
Lubbock AP	33	4	11	15	M
Lufkin AP	31	19	24	28	M
McAllen	26	30	34	38	M
Midland AP	32	13	19	23	M
Mineral Wells AP	32	12	18	22	H
Palestine CO	31	16	21	25	M
Pampa	35	0	7	11	M
Pecos	31	10	15	19	L
Plainview	34	3	10	14	M
Port Arthur AP	30	25	29	33	M
San Angelo, Goodfellow AFB	31	15	20	25	M
San Antonio AP	29	22	25	30	L
Sherman-Perrin AFB	33	12	18	23	H
Snyder	32	9	15	19	M
Temple	31	18	23	27	M
Tyler AP	32	15	20	24	M
Vernon	34	7	14	18	H
Victoria AP	28	24	28	32	M
Waco AP	31	16	21	26	M
Wichita Falls AP	34	9	15	19	H
UTAH					
Cedar City AP	37	−10	−1	6	VL
Logan	41	−7	3	7	VL
Moab	38	2	12	16	VL
Ogden CO	41	−3	7	11	VL
Price	39	−7	3	7	L
Provo	40	−6	2	6	L
Richfield	38	−10	−1	3	L
St. George CO	37	13	22	26	VL
Salt Lake City AP	40	−2	5	9	L
Vernal AP	40	−20	−10	−6	VL
VERMONT					
Barre	44	−23	−17	−13	L
Burlington AP	44	−18	−12	−7	M
Rutland	43	−18	−12	−8	L
VIRGINIA					
Charlottsville	38	7	11	15	L
Danville AP	36	9	13	17	L
Fredericksburg	38	6	10	14	M
Harrisonburg	38	0	5	9	L
Lynchburg AP	37	10	15	19	L
Norfolk AP	36	18	20	23	M
Petersburg	37	10	15	18	L
Richmond AP	37	10	14	18	L
Roanoke AP	37	9	15	18	L
Staunton	38	3	8	12	L
Winchester	39	1	6	10	L
WASHINGTON					
Aberdeen	47	19	24	27	M
Bellingham AP	48	8	14	18	L
Bremerton	47	17	24	29	L
Ellensburg AP	47	−5	2	6	VL
Everett-Paine AFB	47	13	19	24	L
Kennewick	46	4	11	15	VL
Longview	46	14	20	24	L
Moses Lake, Larson AFB	47	−14	−7	−1	VL
Olympia AP	47	15	21	25	L
Port Angeles	48	20	26	29	M
Seattle-Boeing Fld	47	17	23	27	L

State and Station	Lat. °	Median of Annual Extremes	Winter 99%	Winter 97½%	Coincident Wind Velocity
WASHINGTON (continued)					
Seattle CO	47	22	28	32	L
Seattle-Tacoma AP	47	14	20	24	L
Spokane AP	47	− 5	− 2	4	VL
Tacoma-McChord AFB	47	14	20	24	L
Walla Walla AP	46	5	12	16	VL
Wenatchee	47	− 2	5	9	VL
Yakima AP	46	− 1	6	10	VL
WEST VIRGINIA					
Beckley	37	− 4	0	6	L
Bluefield AP	37	1	6	10	L
Charleston AP	38	1	9	14	L
Clarksburg	39	− 2	3	7	L
Elkins AP	38	− 4	1	5	L
Huntington CO	38	4	10	14	L
Martinsburg AP	39	1	6	10	L
Morgantown AP	39	− 2	3	7	L
Parkersburg CO	39	2	8	12	L
Wheeling	40	0	5	9	L
WISCONSIN					
Appleton	44	−16	−10	− 6	M
Ashland	46	−27	−21	−17	L
Beloit	42	−13	− 7	− 3	M
Eau Claire AP	44	−21	−15	−11	L
Fond du Lac	43	−17	−11	− 7	M
Green Bay AP	44	−16	−12	− 7	M
La Crosse AP	43	−18	−12	− 8	M
Madison AP	43	−13	− 9	− 5	M
Manitowoc	44	−11	− 5	− 1	M
Marinette	45	−14	− 8	− 4	M
Milwaukee AP	43	−11	− 6	− 2	M
Racine	42	−10	− 4	0	M
Sheboygan	43	−10	− 4	0	M
Stevens Point	44	−22	−16	−12	M
Waukesha	43	−12	− 6	− 2	M
Wausau AP	44	−24	−18	−14	M
WYOMING					
Casper AP	42	−20	−11	− 5	L
Cheyenne AP	41	−15	− 6	− 2	M
Cody AP	44	−23	−13	− 9	L
Evanston	41	−22	−12	− 8	VL
Lander AP	42	−26	−16	−12	VL
Laramie AP	41	−17	− 6	− 2	M
Newcastle	43	−18	− 9	− 5	M
Rawlins	41	−24	−15	−11	L
Rock Springs AP	41	−16	− 6	− 1	VL
Sheridan AP	44	−21	−12	− 7	L
Torrington	42	−20	−11	− 7	M

Average Monthly and Yearly Degree-Days for the United States[a,b,c]

State	Station	Avg. Winter Temp	July	Aug.	Sept.	Oct.	Nov.	Dec.	Jan.	Feb.	Mar.	Apr.	May	June	Yearly Total
Ala.	Birmingham..............A	54.2	0	0	6	93	363	555	592	462	363	108	9	0	2551
	Huntsville...............A	51.3	0	0	12	127	426	663	694	557	434	138	19	0	3070
	Mobile..................A	59.9	0	0	0	22	213	357	415	300	211	42	0	0	1560
	Montgomery.............A	55.4	0	0	0	68	330	527	543	417	316	90	0	0	2291
Alaska	Anchorage..............A	23.0	245	291	516	930	1284	1572	1631	1316	1293	879	592	315	10864
	Fairbanks..............A	6.7	171	332	642	1203	1833	2254	2359	1901	1739	1068	555	222	14279
	Juneau.................A	32.1	301	338	483	725	921	1135	1237	1070	1073	810	601	381	9075
	Nome...................A	13.1	481	496	693	1094	1455	1820	1879	1666	1770	1314	930	573	14171
Ariz.	Flagstaff...............A	35.6	46	68	201	558	867	1073	1169	991	911	651	437	180	7152
	Phoenix................A	58.5	0	0	0	22	234	415	474	328	217	75	0	0	1765
	Tucson.................A	58.1	0	0	0	25	231	406	471	344	242	75	6	0	1800
	Winslow................A	43.0	0	0	6	245	711	1008	1054	770	601	291	96	0	4782
	Yuma...................A	64.2	0	0	0	0	108	264	307	190	90	15	0	0	974
Ark.	Fort Smith..............A	50.3	0	0	12	127	450	704	781	596	456	144	22	0	3292
	Little Rock.............A	50.5	0	0	9	127	465	716	756	577	434	126	9	0	3219
	Texarkana..............A	54.2	0	0	0	78	345	561	626	468	350	105	0	0	2533
Calif.	Bakersfield.............A	55.4	0	0	0	37	282	502	546	364	267	105	19	0	2122
	Bishop.................A	46.0	0	0	48	260	576	797	874	680	555	306	143	36	4275
	Blue Canyon............A	42.2	28	37	108	347	594	781	896	795	806	597	412	195	5596
	Burbank................A	58.6	0	0	6	43	177	301	366	277	239	138	81	18	1646
	Eureka.................C	49.9	270	257	258	329	414	499	546	470	505	438	372	285	4643
	Fresno.................A	53.3	0	0	0	84	354	577	605	426	335	162	62	6	2611
	Long Beach.............A	57.8	0	0	9	47	171	316	397	311	264	171	93	24	1803
	Los Angeles............A	57.4	28	28	42	78	180	291	372	302	288	219	158	81	2061
	Los Angeles............C	60.3	0	0	6	31	132	229	310	230	202	123	68	18	1349
	Mt. Shasta.............C	41.2	25	34	123	406	696	902	983	784	738	525	347	159	5722
	Oakland................A	53.5	53	50	45	127	309	481	527	400	353	255	180	90	2870
	Red Bluff..............A	53.8	0	0	0	53	318	555	605	428	341	168	47	0	2515
	Sacramento.............A	53.9	0	0	0	56	321	546	583	414	332	178	72	0	2502
	Sacramento.............C	54.4	0	0	0	62	312	533	561	392	310	173	76	0	2419
	Sandberg...............C	46.8	0	0	30	202	480	691	778	661	620	426	264	57	4209
	San Diego..............A	59.5	9	0	21	43	135	236	298	235	214	135	90	42	1458
	San Francisco..........A	53.4	81	78	60	143	306	462	508	395	363	279	214	126	3015
	San Francisco..........C	55.1	192	174	102	118	231	388	443	336	319	279	239	180	3001
	Santa Maria............A	54.3	99	93	96	146	270	391	459	370	363	282	233	165	2967
Colo.	Alamosa................A	29.7	65	99	279	639	1065	1420	1476	1162	1020	696	440	168	8529
	Colorado Springs.......A	37.3	9	25	132	456	825	1032	1128	938	893	582	319	84	6423
	Denver.................A	37.6	6	9	117	428	819	1035	1132	938	887	558	288	66	6283
	Denver.................C	40.8	0	0	90	366	714	905	1004	851	800	492	254	48	5524
	Grand Junction.........A	39.3	0	0	30	313	786	1113	1209	907	729	387	146	21	5641
	Pueblo.................A	40.4	0	0	54	326	750	986	1085	871	772	429	174	15	5462
Conn.	Bridgeport.............A	39.9	0	0	66	307	615	986	1079	966	853	510	208	27	5617
	Hartford...............A	37.3	0	12	117	394	714	1101	1190	1042	908	519	205	33	6235
	New Haven.............A	39.0	0	12	87	347	648	1011	1097	991	871	543	245	45	5897
Del.	Wilmington.............A	42.5	0	0	51	270	588	927	980	874	735	387	112	6	4930
D. C.	Washington.............A	45.7	0	0	33	217	519	834	871	762	626	288	74	0	4224
Fla.	Apalachicola...........C	61.2	0	0	0	16	153	319	347	260	180	33	0	0	1308
	Daytona Beach..........A	64.5	0	0	0	0	75	211	248	190	140	15	0	0	879
	Fort Myers.............A	68.6	0	0	0	0	24	109	146	101	62	0	0	0	442
	Jacksonville...........A	61.9	0	0	0	12	144	310	332	246	174	21	0	0	1239
	Key West...............A	73.1	0	0	0	0	0	28	40	31	9	0	0	0	108
	Lakeland...............C	66.7	0	0	0	0	57	164	195	146	99	0	0	0	661
	Miami..................A	71.1	0	0	0	0	0	65	74	56	19	0	0	0	214

Notes: a. From ASHRAE Guide and Data Book; base temperature 65°F.
 b. A indicates airport; C indicates city.
 c. Average winter temperatures for October through April, inclusive.

Appendix 4B—Continued

State	Station	Avg. Winter Temp	July	Aug.	Sept.	Oct.	Nov.	Dec.	Jan.	Feb.	Mar.	Apr.	May	June	Yearly Total
Fla. (Cont'd)	Miami Beach.........C	72.5	0	0	0	0	0	40	56	36	9	0	0	0	141
	Orlando............A	65.7	0	0	0	0	72	198	220	165	105	6	0	0	766
	Pensacola..........A	60.4	0	0	0	19	195	353	400	277	183	36	0	0	1463
	Tallahassee........A	60.1	0	0	0	28	198	360	375	286	202	36	0	0	1485
	Tampa..............A	66.4	0	0	0	0	60	171	202	148	102	0	0	0	683
	West Palm Beach....A	68.4	0	0	0	0	6	65	87	64	31	0	0	0	253
Ga.	Athens.............A	51.8	0	0	12	115	405	632	642	529	431	141	22	0	2929
	Atlanta............A	51.7	0	0	18	124	417	648	636	518	428	147	25	0	2961
	Augusta............A	54.5	0	0	0	78	333	552	549	445	350	90	0	0	2397
	Columbus...........A	54.8	0	0	0	87	333	543	552	434	338	96	0	0	2383
	Macon..............A	56.2	0	0	0	71	297	502	505	403	295	63	0	0	2136
	Rome...............A	49.9	0	0	24	161	474	701	710	577	468	177	34	0	3326
	Savannah...........A	57.8	0	0	0	47	246	437	437	353	254	45	0	0	1819
	Thomasville........C	60.0	0	0	0	25	198	366	394	305	208	33	0	0	1529
Hawaii	Lihue..............A	72.7	0	0	0	0	0	0	0	0	0	0	0	0	0
	Honolulu...........A	74.2	0	0	0	0	0	0	0	0	0	0	0	0	0
	Hilo...............A	71.9	0	0	0	0	0	0	0	0	0	0	0	0	0
Idaho	Boise..............A	39.7	0	0	132	415	792	1017	1113	854	722	438	245	81	5809
	Lewiston...........A	41.0	0	0	123	403	756	933	1063	815	694	426	239	90	5542
	Pocatello..........A	34.8	0	0	172	493	900	1166	1324	1058	905	555	319	141	7033
Ill.	Cairo..............C	47.9	0	0	36	164	513	791	856	680	539	195	47	0	3821
	Chicago (O'Hare)...A	35.8	0	12	117	381	807	1166	1265	1086	939	534	260	72	6639
	Chicago (Midway)...A	37.5	0	0	81	326	753	1113	1209	1044	890	480	211	48	6155
	Chicago............C	38.9	0	0	66	279	705	1051	1150	1000	868	489	226	48	5882
	Moline.............A	36.4	0	9	99	335	774	1181	1314	1100	918	450	189	39	6408
	Peoria.............A	38.1	0	6	87	326	759	1113	1218	1025	849	426	183	33	6025
	Rockford...........A	34.8	6	9	114	400	837	1221	1333	1137	961	516	236	60	6830
	Springfield........A	40.6	0	0	72	291	696	1023	1135	935	769	354	136	18	5429
Ind.	Evansville.........A	45.0	0	0	66	220	606	896	955	767	620	237	68	0	4435
	Fort Wayne.........A	37.3	0	9	105	378	783	1135	1178	1028	890	471	189	39	6205
	Indianapolis.......A	39.6	0	0	90	316	723	1051	1113	949	809	432	177	39	5699
	South Bend.........A	36.6	0	6	111	372	777	1125	1221	1070	933	525	239	60	6439
Iowa	Burlington.........A	37.6	0	0	93	322	768	1135	1259	1042	859	426	177	33	6114
	Des Moines.........A	35.5	0	6	96	363	828	1225	1370	1137	915	438	180	30	6588
	Dubuque............A	32.7	12	31	156	450	906	1287	1420	1204	1026	546	260	78	7376
	Sioux City.........A	34.0	0	9	108	369	867	1240	1435	1198	989	483	214	39	6951
	Waterloo...........A	32.6	12	19	138	428	909	1296	1460	1221	1023	531	229	54	7320
Kans.	Concordia..........A	40.4	0	0	57	276	705	1023	1163	935	781	372	149	18	5479
	Dodge City.........A	42.5	0	0	33	251	666	939	1051	840	719	354	124	9	4986
	Goodland...........A	37.8	0	6	81	381	810	1073	1166	955	884	507	236	42	6141
	Topeka.............A	41.7	0	0	57	270	672	980	1122	893	722	330	124	12	5182
	Wichita............A	44.2	0	0	33	229	618	905	1023	804	645	270	87	6	4620
Ky.	Covington..........A	41.4	0	0	75	291	669	983	1035	893	756	390	149	24	5265
	Lexington..........A	43.8	0	0	54	239	609	902	946	818	685	325	105	0	4683
	Louisville.........A	44.0	0	0	54	248	609	890	930	818	682	315	105	9	4660
La.	Alexandria.........A	57.5	0	0	0	56	273	431	471	361	260	69	0	0	1921
	Baton Rouge........A	59.8	0	0	0	31	216	369	409	294	208	33	0	0	1560
	Lake Charles.......A	60.5	0	0	0	19	210	341	381	274	195	39	0	0	1459
	New Orleans........A	61.0	0	0	0	19	192	322	363	258	192	39	0	0	1385
	New Orleans........C	61.8	0	0	0	0	165	291	344	241	177	24	0	0	1254
	Shreveport.........A	56.2	0	0	0	47	297	477	552	426	304	81	0	0	2184
Me.	Caribou............A	24.4	78	115	336	682	1044	1535	1690	1470	1308	858	468	183	9767
	Portland...........A	33.0	12	53	195	508	807	1215	1339	1182	1042	675	372	111	7511
Md.	Baltimore..........A	43.7	0	0	48	264	585	905	936	820	679	327	90	0	4654
	Baltimore..........C	46.2	0	0	27	189	486	806	859	762	629	288	65	0	4111
	Frederich..........A	42.0	0	0	66	307	624	955	995	876	741	384	127	12	5087
Mass.	Boston.............A	40.0	0	9	60	316	603	983	1088	972	846	513	208	36	5634
	Nantucket..........A	40.2	12	22	93	332	573	896	992	941	896	621	384	129	5891
	Pittsfield.........A	32.6	25	59	219	524	831	1231	1339	1196	1063	660	326	105	7578
	Worcester..........A	34.7	6	34	147	450	774	1172	1271	1123	998	612	304	78	6969

State	Station	Avg. Winter Temp	July	Aug.	Sept.	Oct.	Nov.	Dec.	Jan.	Feb.	Mar.	Apr.	May	June	Yearly Total
Mich.	Alpena................A	29.7	68	105	273	580	912	1268	1404	1299	1218	777	446	156	8506
	Detroit (City)..........A	37.2	0	0	87	360	738	1088	1181	1058	936	522	220	42	6232
	Detroit (Wayne)........A	37.1	0	0	96	353	738	1088	1194	1061	933	534	239	57	6293
	Detroit (Willow Run)....A	37.2	0	0	90	357	750	1104	1190	1053	921	519	229	45	6258
	Escanaba..............C	29.6	59	87	243	539	924	1293	1445	1296	1203	777	456	159	8481
	Flint..................A	33.1	16	40	159	465	843	1212	1330	1198	1066	639	319	90	7377
	Grand Rapids..........A	34.9	9	28	135	434	804	1147	1259	1134	1011	579	279	75	6894
	Lansing...............A	34.8	6	22	138	431	813	1163	1262	1142	1011	579	273	69	6909
	Marquette.............C	30.2	59	81	240	527	936	1268	1411	1268	1187	771	468	177	8393
	Muskegon..............A	36.0	12	28	120	400	762	1088	1209	1100	995	594	310	78	6696
	Sault Ste. Marie........A	27.7	96	105	279	580	951	1367	1525	1380	1277	810	477	201	9048
Minn.	Duluth................A	23.4	71	109	330	632	1131	1581	1745	1518	1355	840	490	198	10000
	Minneapolis............A	28.3	22	31	189	505	1014	1454	1631	1380	1166	621	288	81	8382
	Rochester..............A	28.8	25	34	186	474	1005	1438	1593	1366	1150	630	301	93	8295
Miss.	Jackson...............A	55.7	0	0	0	65	315	502	546	414	310	87	0	0	2239
	Meridian..............A	55.4	0	0	0	81	339	518	543	417	310	81	0	0	2289
	Vicksburg.............C	56.9	0	0	0	53	279	462	512	384	282	69	0	0	2041
Mo.	Columbia..............A	42.3	0	0	54	251	651	967	1076	874	716	324	121	12	5046
	Kansas City............A	43.9	0	0	39	220	612	905	1032	818	682	294	109	0	4711
	St. Joseph.............A	40.3	0	6	60	285	708	1039	1172	949	769	348	133	15	5484
	St. Louis..............A	43.1	0	0	60	251	627	936	1026	848	704	312	121	15	4900
	St. Louis..............C	44.8	0	0	36	202	576	884	977	801	651	270	87	0	4484
	Springfield.............A	44.5	0	0	45	223	600	877	973	781	660	291	105	6	4900
Mont.	Billings...............A	34.5	6	15	186	487	897	1135	1296	1100	970	570	285	102	7049
	Glasgow...............A	26.4	31	47	270	608	1104	1466	1711	1439	1187	648	335	150	8996
	Great Falls.............A	32.8	28	53	258	543	921	1169	1349	1154	1063	642	384	186	7750
	Havre.................A	28.1	28	53	306	595	1065	1367	1584	1364	1181	657	338	162	8700
	Havre.................C	29.8	19	37	252	539	1014	1321	1528	1305	1116	612	304	135	8182
	Helena................A	31.1	31	59	294	601	1002	1265	1438	1170	1042	651	381	195	8129
	Kalispell...............A	31.4	50	99	321	654	1020	1240	1401	1134	1029	639	397	207	8191
	Miles City.............A	31.2	6	6	174	502	972	1296	1504	1252	1057	579	276	99	7723
	Missoula..............A	31.5	34	74	303	651	1035	1287	1420	1120	970	621	391	219	8125
Neb.	Grand Island...........A	36.0	0	6	108	381	834	1172	1314	1089	908	462	211	45	6530
	Lincoln................C	38.8	0	6	75	301	726	1066	1237	1016	834	402	171	30	5864
	Norfolk................A	34.0	9	0	111	397	873	1234	1414	1179	983	498	233	48	6979
	North Platte...........A	35.5	0	6	123	440	885	1166	1271	1039	930	519	248	57	6684
	Omaha................A	35.6	0	12	105	357	828	1175	1355	1126	939	465	208	42	6612
	Scottsbluff.............A	35.9	0	0	138	459	876	1128	1231	1008	921	552	285	75	6673
	Valentine..............A	32.6	9	12	165	493	942	1237	1395	1176	1045	579	288	84	7425
Nev.	Elko..................A	34.0	9	34	225	561	924	1197	1314	1036	911	621	409	192	7433
	Ely...................A	33.1	28	43	234	592	939	1184	1308	1075	977	672	456	225	7733
	Las Vegas..............A	53.5	0	0	0	78	387	617	688	487	335	111	6	0	2709
	Reno..................A	39.3	43	87	204	490	801	1026	1073	823	729	510	357	189	6332
	Winnemucca...........A	36.7	0	34	210	536	876	1091	1172	916	837	573	363	153	6761
N. H.	Concord...............A	33.0	6	50	177	505	822	1240	1358	1184	1032	636	298	75	7383
	Mt. Washington Obsv....	15.2	493	536	720	1057	1341	1742	1820	1663	1652	1260	930	603	13817
N. J.	Atlantic City...........A	43.2	0	0	39	251	549	880	936	848	741	420	133	15	4812
	Newark...............A	42.8	0	0	30	248	573	921	983	876	729	381	118	0	4589
	Trenton...............C	42.4	0	0	57	264	576	924	989	885	753	399	121	12	4980
N. M.	Albuquerque...........A	45.0	0	0	12	229	642	868	930	703	595	288	81	0	4348
	Clayton...............A	42.0	0	6	66	310	699	899	986	812	747	429	183	21	5158
	Raton.................A	38.1	9	28	126	431	825	1048	1116	904	834	543	301	63	6228
	Roswell...............A	47.5	0	0	18	202	573	806	840	641	481	201	31	0	3793
	Silver City.............A	48.0	0	0	6	183	525	729	791	605	518	261	87	0	3705
N. Y.	Albany................A	34.6	0	19	138	440	777	1194	1311	1156	992	564	239	45	6875
	Albany................C	37.2	0	9	102	375	699	1104	1218	1072	908	498	186	30	6201
	Binghamton...........A	33.9	22	65	201	471	810	1184	1277	1154	1045	645	313	99	7286
	Binghamton...........C	36.6	0	28	141	406	732	1107	1190	1081	949	543	229	45	6451
	Buffalo................A	34.5	19	37	141	440	777	1156	1256	1145	1039	645	329	78	7062
	New York (Cent. Park)..	42.8	0	0	30	233	540	902	986	885	760	408	118	9	4871
	New York (La Guardia)..A	43.1	0	0	27	223	528	887	973	879	750	414	124	6	4811

State	Station	Avg. Winter Temp	July	Aug.	Sept.	Oct.	Nov.	Dec.	Jan.	Feb.	Mar.	Apr.	May	June	Yearly Total
	New York (Kennedy)....A	41.4	0	0	36	248	564	933	1029	935	815	480	167	12	5219
	Rochester.............A	35.4	9	31	126	415	747	1125	1234	1123	1014	597	279	48	6748
	Schenectady...........C	35.4	0	22	123	422	756	1159	1283	1131	970	543	211	30	6650
	Syracuse..............A	35.2	6	28	132	415	744	1153	1271	1140	1004	570	248	45	6756
N. C.	Asheville.............C	46.7	0	0	48	245	555	775	784	683	592	273	87	0	4042
	Cape Hatteras..........	53.3	0	0	0	78	273	521	580	518	440	177	25	0	2612
	Charlotte.............A	50.4	0	0	6	124	438	691	691	582	481	156	22	0	3191
	Greensboro............A	47.5	0	0	33	192	513	778	784	672	552	234	47	0	3805
	Raleigh...............A	49.4	0	0	21	164	450	716	725	616	487	180	34	0	3393
	Wilmington............A	54.6	0	0	0	74	291	521	546	462	357	96	0	0	2347
	Winston-Salem.........A	48.4	0	0	21	171	483	747	753	652	524	207	37	0	3595
N. D.	Bismarck..............A	26.6	34	28	222	577	1083	1463	1708	1442	1203	645	329	117	8851
	Devils Lake...........C	22.4	40	53	273	642	1191	1634	1872	1579	1345	753	381	138	9901
	Fargo.................A	24.8	28	37	219	574	1107	1569	1789	1520	1262	690	332	99	9226
	Williston.............A	25.2	31	43	261	601	1122	1513	1758	1473	1262	681	357	141	9243
Ohio	Akron-Canton..........A	38.1	0	9	96	381	726	1070	1138	1016	871	489	202	39	6037
	Cincinnati............C	45.1	0	0	39	208	558	862	915	790	642	294	96	6	4410
	Cleveland.............A	37.2	9	25	105	384	738	1088	1159	1047	918	552	260	66	6351
	Columbus..............A	39.7	0	6	84	347	714	1039	1088	949	809	426	171	27	5660
	Columbus..............C	41.5	0	0	57	285	651	977	1032	902	760	396	136	15	5211
	Dayton................A	39.8	0	6	78	310	696	1045	1097	955	809	429	167	30	5622
	Mansfield.............A	36.9	9	22	114	397	768	1110	1169	1042	924	543	245	60	6403
	Sandusky..............C	39.1	0	6	66	313	684	1032	1107	991	868	495	198	36	5796
	Toledo................A	36.4	0	16	117	406	792	1138	1200	1056	924	543	242	60	6494
	Youngstown............A	36.8	6	19	120	412	771	1104	1169	1047	921	540	248	60	6417
Okla.	Oklahoma City.........A	48.3	0	0	15	164	498	766	868	664	527	189	34	0	3725
	Tulsa.................A	47.7	0	0	18	158	522	787	893	683	539	213	47	0	3860
Ore.	Astoria...............A	45.6	146	130	210	375	561	679	753	622	636	480	363	231	5186
	Burns.................C	35.9	12	37	210	515	867	1113	1246	988	856	570	366	177	6957
	Eugene................A	45.6	34	34	129	366	585	719	803	627	589	426	279	135	4726
	Meacham...............A	34.2	84	124	288	580	918	1091	1209	1005	983	726	527	339	7874
	Medford...............A	43.2	0	0	78	372	678	871	918	697	642	432	242	78	5008
	Pendleton.............A	42.6	0	0	111	350	711	884	1017	773	617	396	205	63	5127
	Portland..............A	45.6	25	28	114	335	597	735	825	644	586	396	245	105	4635
	Portland..............C	47.4	12	16	75	267	534	679	769	594	536	351	198	78	4109
	Roseburg..............A	46.3	22	16	105	329	567	713	766	608	570	405	267	123	4491
	Salem.................A	45.4	37	31	111	338	594	729	822	647	611	417	273	144	4754
Pa.	Allentown.............A	38.9	0	0	90	353	693	1045	1116	1002	849	471	167	24	5810
	Erie..................A	36.8	0	25	102	391	714	1063	1169	1081	973	585	288	60	6451
	Harrisburg............A	41.2	0	0	63	298	648	992	1045	907	766	396	124	12	5251
	Philadelphia..........A	41.8	0	0	60	297	620	965	1016	889	747	392	118	40	5144
	Philadelphia..........C	44.5	0	0	30	205	513	856	924	823	691	351	93	0	4486
	Pittsburgh............A	38.4	0	9	105	375	726	1063	1119	1002	874	480	195	39	5987
	Pittsburgh............C	42.2	0	0	60	291	615	930	983	885	763	390	124	12	5053
	Reading...............C	42.4	0	0	54	257	597	939	1001	885	735	372	105	0	4945
	Scranton..............A	37.2	0	19	132	434	762	1104	1156	1028	893	498	195	33	6254
	Williamsport..........A	38.5	0	9	111	375	717	1073	1122	1002	856	468	177	24	5934
R. I.	Block Island..........A	40.1	0	16	78	307	594	902	1020	955	877	612	344	99	5804
	Providence............A	38.8	0	16	96	372	660	1023	1110	988	868	534	236	51	5954
S. C.	Charleston............A	56.4	0	0	0	59	282	471	487	389	291	54	0	0	2033
	Charleston............C	57.9	0	0	0	34	210	425	443	367	273	42	0	0	1794
	Columbia..............A	54.0	0	0	0	84	345	577	570	470	357	81	0	0	2484
	Florence..............A	54.5	0	0	0	78	315	552	552	459	347	84	0	0	2387
	Greenville-Spartenburg...A	51.6	0	0	6	121	399	651	660	546	446	132	19	0	2980
S. D.	Huron.................A	28.8	9	12	165	508	1014	1432	1628	1355	1125	600	288	87	8223
	Rapid City............A	33.4	22	12	165	481	897	1172	1333	1145	1051	615	326	126	7345
	Sioux Falls...........A	30.6	19	25	168	462	972	1361	1544	1285	1082	573	270	78	7839
Tenn.	Bristol...............A	46.2	0	0	51	236	573	828	828	700	598	261	68	0	4143
	Chattanooga...........A	50.3	0	0	18	143	468	698	722	577	453	150	25	0	3254
	Knoxville.............A	49.2	0	0	30	171	489	725	732	613	493	198	43	0	3494
	Memphis...............A	50.5	0	0	18	130	447	698	729	585	456	147	22	0	3232

State or Prov.	Station	Avg. Winter Temp	July	Aug.	Sept.	Oct.	Nov.	Dec.	Jan.	Feb.	Mar.	Apr.	May	June	Yearly Total
	Memphis...............C	51.6	0	0	12	102	396	648	710	568	434	129	16	0	3015
	Nashville..............A	48.9	0	0	30	158	495	732	778	644	512	189	40	0	3578
	Oak Ridge.............C	47.7	0	0	39	192	531	772	778	669	552	228	56	0	3817
Tex.	Abilene................A	53.9	0	0	0	99	366	586	642	470	347	114	0	0	2624
	Amarillo...............A	47.0	0	0	18	205	570	797	877	664	546	252	56	0	3985
	Austin.................A	59.1	0	0	0	31	225	388	468	325	223	51	0	0	1711
	Brownsville............A	67.7	0	0	0	0	66	149	205	106	74	0	0	0	600
	Corpus Christi.........A	64.6	0	0	0	0	120	220	291	174	109	0	0	0	914
	Dallas.................A	55.3	0	0	0	62	321	524	601	440	319	90	6	0	2363
	El Paso................A	52.9	0	0	0	84	414	648	685	445	319	105	0	0	2700
	Fort Worth............A	55.1	0	0	0	65	324	536	614	448	319	99	0	0	2405
	Galveston..............A	62.2	0	0	0	6	147	276	360	263	189	33	0	0	1274
	Galveston..............C	62.0	0	0	0	0	138	270	350	258	189	30	0	0	1235
	Houston................A	61.0	0	0	0	6	183	307	384	288	192	36	0	0	1396
	Houston................C	62.0	0	0	0	0	165	288	363	258	174	30	0	0	1278
	Laredo.................A	66.0	0	0	0	0	105	217	267	134	74	0	0	0	797
	Lubbock................A	48.8	0	0	18	174	513	744	800	613	484	201	31	0	3578
	Midland................A	53.8	0	0	0	87	381	592	651	468	322	90	0	0	2591
	Port Arthur............A	60.5	0	0	0	22	207	329	384	274	192	39	0	0	1447
	San Angelo.............A	56.0	0	0	0	68	318	536	567	412	288	66	0	0	2255
	San Antonio............A	60.1	0	0	0	31	204	363	428	286	195	39	0	0	1546
	Victoria...............A	62.7	0	0	0	6	150	270	344	230	152	21	0	0	1173
	Waco...................A	57.2	0	0	0	43	270	456	536	389	270	66	0	0	2030
	Wichita Falls..........A	53.0	0	0	0	99	381	632	698	518	378	120	6	0	2832
Utah	Milford................A	36.5	0	0	99	443	867	1141	1252	988	822	519	279	87	6497
	Salt Lake City.........A	38.4	0	0	81	419	849	1082	1172	910	763	459	233	84	6052
	Wendover...............A	39.1	0	0	48	372	822	1091	1178	902	729	408	177	51	5778
Vt.	Burlington.............A	29.4	28	65	207	539	891	1349	1513	1333	1187	714	353	90	8269
Va.	Cape Henry............C	50.0	0	0	0	112	360	645	694	633	536	246	53	0	3279
	Lynchburg..............A	46.0	0	0	51	223	540	822	849	731	605	267	78	0	4166
	Norfolk................A	49.2	0	0	0	136	408	698	738	655	533	216	37	0	3421
	Richmond...............A	47.3	0	0	36	214	495	784	815	703	546	219	53	0	3865
	Roanoke................A	46.1	0	0	51	229	549	825	834	722	614	261	65	0	4150
Wash.	Olympia................A	44.2	68	71	198	422	636	753	834	675	645	450	307	177	5236
	Seattle-Tacoma.........A	44.2	56	62	162	391	633	750	828	678	657	474	295	159	5145
	Seattle................C	46.9	50	47	129	329	543	657	738	599	577	396	242	117	4424
	Spokane................A	36.5	9	25	168	493	879	1082	1231	980	834	531	288	135	6655
	Walla Walla............C	43.8	0	0	87	310	681	843	986	745	589	342	177	45	4805
	Yakima.................A	39.1	0	12	144	450	828	1039	1163	868	713	435	220	69	5941
W. Va.	Charleston.............A	44.8	0	0	63	254	591	865	880	770	648	300	96	9	4476
	Elkins.................A	40.1	9	25	135	400	729	992	1008	896	791	444	198	48	5675
	Huntington.............A	45.0	0	0	63	257	585	856	880	764	636	294	99	12	4446
	Parkersburg............C	43.5	0	0	60	264	606	905	942	826	691	339	115	6	4754
Wisc.	Green Bay..............A	30.3	28	50	174	484	924	1333	1494	1313	1141	654	335	99	8029
	La Crosse..............A	31.5	12	19	153	437	924	1339	1504	1277	1070	540	245	69	7589
	Madison................A	30.9	25	40	174	474	930	1330	1473	1274	1113	618	310	102	7863
	Milwaukee..............A	32.6	43	47	174	471	876	1252	1376	1193	1054	642	372	135	7635
Wyo.	Casper.................A	33.4	6	16	192	524	942	1169	1290	1084	1020	657	381	129	7410
	Cheyenne...............A	34.2	28	37	219	543	909	1085	1212	1042	1026	702	428	150	7381
	Lander.................A	31.4	6	19	204	555	1020	1299	1417	1145	1017	654	381	153	7870
	Sheridan...............A	32.5	25	31	219	539	948	1200	1355	1154	1031	642	366	150	7680

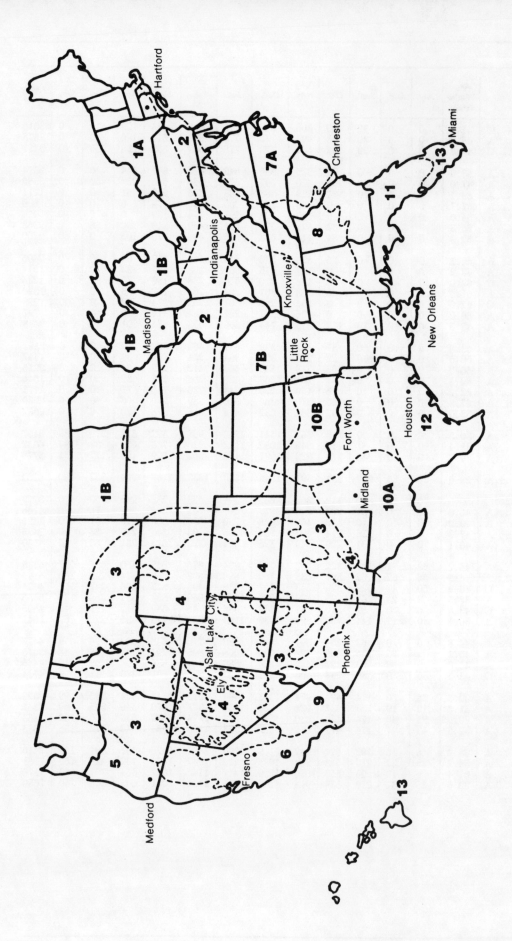

†The 13 general climate regions have been developed from ongoing research at the AIA Research Corporation in conjunction with the National Climatic Center of NOAA. The regions have been based on heating and cooling needs, solar usefulness in a 50° to 60°F range, wind usefulness in a 75° to 85°F range, diurnal temperature impact, and low humidity impact for natural heating and cooling of homes. Further information is available in Technical Report No. 1 from the AIA/RC.

Summary: Some Typical Cities

CLIMATE	CITY WITH PROTOTYPICAL CLIMATE	BASIC CLIMATE CONDITION % CONDITIONING PER YEAR BASED ON TEMPERATURE AND HUMIDITY ALONE	DESIGN PRIORITIES CLIMATIC LIABILITIES (TEMP / WIND / MOISTURE / SUN)	CLIMATIC ASSETS (TEMP / WIND / MOISTURE / SUN)	POTENTIAL CLIMATE CONDITION % CONDITIONING PER YEAR BASED ON SIMPLE BUILDING RESPONSES TO CLIMATE. AT LEAST:	% IMPROVED COMFORT AT LEAST
1A	HARTFORD, CT	13% TOO HOT FOR COMFORT / 75% TOO COOL FOR COMFORT	— / 2 / — / 4	— / 5 / — / 3	0% TOO HOT FOR COMFORT / 63% TOO COOL FOR COMFORT	37%
1B	MADISON, WI	12% TOO HOT FOR COMFORT / 76% TOO COOL FOR COMFORT	1 / 2 / — / 5	4 / 6 / — / 3	7% TOO HOT FOR COMFORT / 65% TOO COOL FOR COMFORT	28%
2	INDIANAPOLIS, IN	20% TOO HOT FOR COMFORT / 66% TOO COOL FOR COMFORT	4 / 2 / — / 5	— / 6 / — / 3	9% TOO HOT FOR COMFORT / 59% TOO COOL FOR COMFORT	32%
3	SALT LAKE CITY, UT	11% TOO HOT FOR COMFORT / 77% TOO COOL FOR COMFORT	1 / 2 / — / 6	3 / 5 / 4 / 1	1% TOO HOT FOR COMFORT / 61% TOO COOL FOR COMFORT	38%
4	ELY, NV	0% TOO HOT FOR COMFORT / 92% TOO COOL FOR COMFORT	1 / 3 / — / —	4 / — / — / 2	0% TOO HOT FOR COMFORT / 76% TOO COOL FOR COMFORT	24%
5	MEDFORD, OR	8% TOO HOT FOR COMFORT / 79% TOO COOL FOR COMFORT	— / 3 / — / 4	3 / 5 / 6 / 2	0% TOO HOT FOR COMFORT / 66% TOO COOL FOR COMFORT	34%
6	FRESNO, CA	17% TOO HOT FOR COMFORT / 62% TOO COOL FOR COMFORT	1 / 7 / 6 / 4	3 / — / — / 1	5% TOO HOT FOR COMFORT / 43% TOO COOL FOR COMFORT	52%
7A	CHARLESTON, SC	42% TOO HOT FOR COMFORT / 46% TOO COOL FOR COMFORT	2 / 5 / — / —	— / 2 / — / 3	33% TOO HOT FOR COMFORT / 33% TOO COOL FOR COMFORT	34%

CLIMATE	CITY WITH PROTOTYPICAL CLIMATE	BASIC CLIMATE CONDITION % CONDITIONING PER YEAR BASED ON TEMPERATURE AND HUMIDITY ALONE	DESIGN PRIORITIES — CLIMATIC LIABILITIES (TEMP / WIND / MOISTURE / SUN)	DESIGN PRIORITIES — CLIMATIC ASSETS (TEMP / WIND / MOISTURE / SUN)	POTENTIAL CLIMATE CONDITION % CONDITIONING PER YEAR BASED ON SIMPLE BUILDING RESPONSES TO CLIMATE, AT LEAST:	% IMPROVED COMFORT AT LEAST
7B	LITTLE ROCK, AR	35% TOO HOT FOR COMFORT / 52% TOO COOL FOR COMFORT	1 / — / 6 / 5 ; 1 / 4 / — / —	— / 3 / — / — ; — / — / — / 2	24% TOO HOT FOR COMFORT / 42% TOO COOL FOR COMFORT	34%
8	KNOXVILLE, TN	28% TOO HOT FOR COMFORT / 56% TOO COOL FOR COMFORT	— / — / 6 / 4 ; 1 / 5 / — / —	2 / 3 / — / — ; 2 / — / — / 4	19% TOO HOT FOR COMFORT / 45% TOO COOL FOR COMFORT	36%
9	PHOENIX, AZ	37% TOO HOT FOR COMFORT / 48% TOO COOL FOR COMFORT	1 / — / — / 2 ; 5 / — / — / —	4 / 7 / 3 / — ; 4 / — / — / 6	18% TOO HOT FOR COMFORT / 30% TOO COOL FOR COMFORT	52%
10A	MIDLAND, TX	26% TOO HOT FOR COMFORT / 55% TOO COOL FOR COMFORT	4 / — / — / 3 ; 4 / 5 / — / —	7 / 6 / 1 / — ; 7 / — / — / 2	10% TOO HOT FOR COMFORT / 37% TOO COOL FOR COMFORT	53%
10B	FORT WORTH, TX	39% TOO HOT FOR COMFORT / 47% TOO COOL FOR COMFORT	— / — / 6 / 3 ; — / 4 / — / —	5 / 1 / 6 / — ; 5 / — / — / 2	24% TOO HOT FOR COMFORT / 33% TOO COOL FOR COMFORT	43%
11	NEW ORLEANS, LA	52% TOO HOT FOR COMFORT / 36% TOO COOL FOR COMFORT	— / — / 4 / 2 ; 1 / 6 / — / —	3 / 1 / — / — ; 3 / — / — / 5	39% TOO HOT FOR COMFORT / 25% TOO COOL FOR COMFORT	36%
12	HOUSTON, TX	54% TOO HOT FOR COMFORT / 35% TOO COOL FOR COMFORT	1 / — / 4 / 3 ; 7 / 5 / — / —	3 / 2 / — / — ; — / — / — / 6	43% TOO HOT FOR COMFORT / 25% TOO COOL FOR COMFORT	32%
13	MIAMI, FL	69% TOO HOT FOR COMFORT / 11% TOO COOL FOR COMFORT	6 / — / 4 / 2 ; 6 / — / — / —	— / 3 / — / — ; — / — / 5 / 5	39% TOO HOT FOR COMFORT / 7% TOO COOL FOR COMFORT	54%

Appendix B:
Bibliography

SOLAR

Anderson, Bruce. **Solar Energy and Shelter Design.** Total Environmental Action, Inc., Harrisville, NH 1973 150 pp. $7.00
Discusses energy conservation and ways of using solar energy, with specific projects and many useful references. Used as a college text.

Anderson, Bruce. **Solar Energy: Fundamentals In Building Design.** McGraw-Hill, New York, NY 1977 374 pp. $21.50 (HBK)
Authoritative text with practical information on how to maximize solar energy's utilization in buildings. Geared to professional architectural designers and engineers, but also adaptable to classroom use.

Anderson, Bruce and Michael Riordan. **The Solar Home Book.** Cheshire Books, Harrisville, NH 1976 304 pp. $8.50 (pbk)
A unique, extremely informative, practical book covering energy conservation, passive solar design and active solar systems all in a single, clearly written volume. Both large and small scale projects are discussed. Most comprehensive source to date. Widely used as a college textbook.

Baer, Steve. **Sunspots-Collected Facts and Solar Fiction.**

Behrman, Daniel. **Solar Energy: The Awakening Science.** Little, Brown & Co., Boston, MA 1976 408 pp. $12.50 (or $10.50 from: Wilderness Society Book Service, 40 Guernsey St., Stamford, CT 06004)
History and principles behind solar energy usage. Describes both large and small scale applications, and assesses future possibilities.

Brace Research Institute. **Survey of Solar Agricultural Dryers.** Brace Research Institute, MacDonald College of McGill University, Ste. Anne de Bellevue 800, Quebec, Canada 1975 170 pp. $7.00
Excellent descriptions of 24 different crop, food and lumber dryers. Photos, diagrams and cost estimates included.

Brinkworth, Brian. **Solar Energy for Man.**

Carlson, Bradley. **Solar Primer One.** SOLARC, Whittier, CA 1975 101pp. $8.95
Presents information on basic solar applications and building design. Written by architects.

Crowther, Richard et al. **Sun Earth.** Crowther/Solar Group, Denver, CO 1976 232 pp. $13.00 (pbk)
Considered to be **the** sourcebook for passive solar building design information. Hundreds of illustrations on how the house and site can act as a solar collector.

Daniels, Farrington. **Direct Use of the Sun's Energy.**

Daniels, George. **Solar Homes and Sun Heating.** Harper & Row, New York, NY 1976 178 pp. $8.95
Practical guide to solar heating for the layman. Non-technical descriptions of basic principles, existing systems and techniques for construction and installation.

Dawson, Joe. **Buying Solar.** Federal Energy Administration and H.E.W. Office of Consumer Affairs, available from: Superintendent of Documents, U.S. Government Printing Office, Washington, D.C. 20402 Stock no. 041-018-00120-4 $1.85 June 1976 71 pp.
Presents solar facts and figures important to students as well as consumers. A clearly written guide with good advice and intelligent tips on what to be aware of when considering solar systems and the purchase thereof. Charts, tables and solar radiation maps included.

de Winter, Francis. **How to Design and Build a Solar Swimming Pool Heater.** Copper Development Association, New York, NY 1975 46 pp. Free
Contains useful information on collector design, construction and economics although oriented specifically to copper, flat-plate water heating collectors.

Duffie, John and William Beckman. **Solar Energy Thermal Processes.** Wiley and Sons, New York, NY 1974 386 pp. $19.00
Advanced textbook restricted to thermal processes in which solar radiation is absorbed by a surface and converted to heat. Emphasis is

on solar meeting thermal energy needs in buildings. Primarily for engineers.

EARS.   **Solar Directory.** (see Secondary pg. 8)

Ewers, William.   **Solar Energy: A Biased Guide.** Domus International, Library of Ecology, Northbrook, IL   1977   96 pp. $4.95
Purpose is to increase public awareness of solar potential and its effective utilization. Non-technical approach with clear diagrams of solar applications and an historical outline of solar development.

Fisher, Rick and Bill Yanda.   **The Food and Heat Producing Solar Greenhouse: Design, Construction, Operation.** John Muir Publications, P.O. Box 613, Sante Fe, NM 87501   1976   162 pp. $6.00 (pbk)
Describes the design, construction and operation of a variety of attached solar greenhouses. Illustrated with photos and diagrams of assorted installations as well as sun movement charts. Excellent introduction to the subject.

Fleck, Paul et al.   **Solar Energy Handbook.** Time-Wise Publications, Pasadena, CA   1975   92 pp. $3.95
For the designer and serious experimenter, this small solar facts-book provides scientific and engineering definitions and units to aid in constructing solar devices. Also includes cost equivalents and contacts to solar energy organizations. Understandable and instructive.

Gropp, Louis.   **Solar Houses: 48 Energy-Saving Designs.** Pantheon Books (Random House), New York, NY   1978   160 pp. $8.95 (pbk)
Nation-wide look at new solar homes. Discusses passive, active and underground designs. Includes sections on demonstration houses, retrofitting, insulation, as well as solar terms, legislation and other important items. Vividly illustrated with full-color and b&w photos and diagrams.

Halacy, Daniel Jr.   **The Coming Age of Solar Energy.** (see Secondary pg. 8)

Hayes, Denis.   **Energy: The Solar Prospect.** (see Secondary pg. 8)

Hickok, Floyd.   **Handbook of Solar and Wind Energy.** Cahners Books, Boston, MA   1975   125 pp. $20.00
Explores the development of renewable energy resources, postulating that all solar sources will increasingly be used as entrepenuers discover and exploit them. Provides an overview of solar heating and cooling and related technologies.

Hovel, H.J.   **Solar Cells.** Academic Press   1975   250 pp. $14.50
Technical information on photovoltaics (solar cells).

Joint Venture and Friends.   **Here Comes the Sun, 1981: Solar-Heated Multi-Family Housing.** (see Secondary pg. 8)

Keyes, John.   **Harnessing the Sun to Heat Your House.** (see Secondary pg. 8)

Kreider, Jan and Frank Kreith.   **Solar Heating and Cooling: Engineering, Practical Design and Economics.** McGraw-Hill, New York, NY   1975   342 pp. $24.75
How-to handbook on solar heating and cooling systems with extensive tables, charts and graph materials. Thorough, detailed and meant for engineers.

Kreith, Frank and Jan Kreider.   **Principles of Solar Engineering.** McGraw-Hill, New York, NY   1977   600 pp. $24.50
An engineering text presenting all aspects of solar energy utilization in a teachable framework. Appropriate for college seniors, graduate students and practicing engineers.

Landa, H.C. et al.   **The Solar Energy Handbook.** Film Instruction Company of America, Milwaukee, WI   1977   154 pp. $9.00
A compilation of theory, technical data and practical moderate-size applications. Much of this data has been used in texts and related technical books.

Lucas, Ted.   **How to Use Solar Energy.** Ward Ritchie Press, Pasadena, CA   1977   315 pp. $7.95
Practical, basic guide to building and buying solar collectors. Includes bibliography, glossary and list of manufacturers.

McCullagh, James.   **The Solar Greenhouse Book.** Rodale Press; Emmaus, PA   1978   314 pp. $8.95 (pbk)/$10.95 (hbk)
The most comprehensive book to date on designing and constructing solar greenhouses. Also offers a thorough discussion of crop production and the biological implications of solar greenhouses. Resplendent with photos and diagrams illustrating a wide variety of structures in use all over the country, adapted to every kind of home, climate and need.

McVeigh, J.C. **Sun Power: An Introduction to the Applications of Solar Energy.** Pergamon Press, New York, NY 1977 180 pp. $4.95
Analyzes all aspects of solar energy and its global applications. Includes thermal and small scale applications, photovoltaics, biological conversion systems, photochemistry and wind power.

Mazria, Edward. **The Passive Solar Energy Book.** Rodale Press, Emmaus, PA 1979 Trade: 300-450 pp. $10.95 (pbk) $12.95 (hbk)//Professional: 450-550 pp. $19.95 (hbk)
A workbook discussing many passive solar energy systems. Begins with general solar theory, then proceeds into 27 specific design patterns. Two separate editions will be published by Spring '79; trade copy for the layman and a more technical and expanded version for professionals and architecture students.

Meinel, Adel and Marjorie Meinel. **Applied Solar Energy: An Introduction.** Addison-Wesley, Reading, MA 1976 651 pp. $17.95
Basic textbook Introduction to the theory of solar energy. Intended for college seniors and graduate students.

Merrigan, Joseph. **Sunlight to Electricity—The Prospects for Solar Energy Conversion by Photovoltaics.** Massachusetts Institute Technology Press, Cambridge, MA 1975 163 pp. $12.95
Covers the fundamentals of photovoltaic conversion as well as offering technological forecasts with market analysis and economics to predict future business opportunities.

Morris, David. **Dawning of Solar Cells.** (see Secondary pg. 8)

Norton, Thomas W. **Solar Energy Experiments for High School and College Students.** (see Secondary pg. 8)

Office of Technology Assessment. **Application of Solar Technology to Today's Energy Needs.** OTA, U.S. Congress, available from: Superintendent of Documents, U.S. Government Printing Office, Washington, D.C. 20402 Vol. I: June 1978 525 pp. $7.00 Stock no. 052-003-00539; Vol. II: Sept. 1978 756 pp. $7.00 Stock no. 052-003-00608
Volume I is policy-oriented and presents an overview of major impacts and constraints on solar markets. Summarizes the analysis of system performance and costs, and reviews direct solar technology. Volume II is very technical and composed almost entirely of charts and tables. Discusses analytical methods and systems analysis. Both are definitely geared to the serious student and solar enthusiast. A thorough treatment of the subject.

Palz, W. **Solar Electricity: An Economic Approach to Solar Energy.** Butterworth, Kent, England. 1977 352 pp.
All known ways of converting the sun's radiation into useful power, including technology, cost projections and latest R&D. Assesses future impact of solar on developing and industrialized countries.

Rankins, William and David Wilson. **Practical Sun Power.** (see Secondary pg. 9)

Sands, Jonathon. **Solar Heating Systems.** (see Secondary pg. 9)

Schwartz, Marris Sokol, ed. **Harvesting the Sun's Energy.** Design III Printing, Fullerton, CA 1975 653 pp. $8.00
Compendium of valuable technical information.

Soully, Dan, Don Prowler and Bruce Anderson. **The Fuel Savers: A Kit of Solar Ideas for Existing Homes.** Cheshire Books, Harrisville, NH 1978 60 pp. $2.75
Eighteen affordable ways to use solar energy in the home. Do-it-yourself guide to low-cost solar retrofitting. Cleverly illustrated, handy size.

Shurcliff, William. **Thermal Shutters and Shades.** William Shurcliff, 19 Appleton St., Cambridge, MA 02138 1977 200 pp. $12.00
Over 100 ideas about passive solar heating and reducing heat loss through windows.

Sinha, Evelyn and Bonnie McCosh. **Solar Energy Technology: A State of the Art Bibliography.** (see Secondary pg. 9)

Solar Age editors. **Solar Age Catalog.** (see Secondary pg. 9)

Stromberg, R.P. and S.O. Woodall. **Passive Solar Building: A Compilation of Data and Results.** NTIS, U.S. Dept. of Commerce, Springfield, VA 1977 71 pp. $5.25
Considered the best overview of passive solar heating available to date. Includes design examples and sources of additional information.

Szokolay, S.V. **Solar Energy and Building.** Wiley and Sons, New York, NY 1975 148 pp. $18.50
Survey of solar heating and cooling techniques. Provides a conceptual understanding of the

problems and solutions of solar energy. Contains illustrated architectural review of solar houses including plans and performance data.

TEA, Inc. **Solar Energy Home Design in Four Climates.** Total Environmental Action, Harrisville, NH 1975 200 pp. $12.75
Explains engineering and design methods for solar homes in climatically different regions of the country to illustrate commercial feasibility of solar heating. Extensive tables and references. An architectural background is helpful.

Turner, Rufus P. **Solar Cells and Photocells.** Howard W. Sams & Company, Indianapolis, IN 1975 96 pp. $3.95
Intended for experimenters, technicians and science fair participants. Gives schematics for operating circuits.

Watson, Donald. **Designing and Building a Solar House: Your Place in the Sun.** Garden Way Publishing, Charlotte, VT 1977 282 pp. $8.95
Details for workable, energy-efficient solar homes. Many drawings, graphs and photos clarify passive and active solar heating and cooling systems, and house design. Cost comparison of different systems included.

Williams, J. Richard. **Solar Energy: Technology and Applications.** (see Secondary pg. 9)

Wilson, David. **Creating Energy.** Lorien House, Black Mountain, NC 1978 56 pp. $5.00 (pbk)
A collection of projects including ways to concentrate the sun's energy to make steam, to construct a solar furnace and to experiment with solar cells.

Wright, David. **Natural Solar Architecture: a passive primer.** Litton Educational Publishing, Inc., New York, NY 1978 245 pp. $7.95 (pbk)
Technical and nontechnical information on how to effectively use passive (direct) solar in architecture. Explains basic solar concepts and its uses in heating and cooling homes and buildings. Text, drawings and diagrams bear the mark of a draftsman. Aimed at students and professionals.

Yanda, Bill and Susan. **An Attached Solar Greenhouse.** The Lightning Tree, P.O. Box 1837, Santa Fe, NM 87501 1976 20 pp. $2.00
Well-illustrated manual for an owner-built greenhouse. Takes into account solar angle, insulation, foundation, design, construction and operation. Written in English and Spanish.

Yellott, John and Carl MacPhee. **Solar Energy Utilization for Heating and Cooling.** Total Environmental Action, Harrisville, NH 1974 20 pp. $2.00
An introduction to the technical and engineering aspects of using solar energy.

Appendix C:
Glossary

Absorber: The absorber is that part of the solar collector that receives the solar radiation and transforms it into thermal energy. It usually is a solid surface through which energy is transferred to the transfer medium (air or liquid).

Absorption cooling: Refrigeration or air conditioning achieved by an absorption-description process that can utilize solar heat to produce a cooling effect.

Absorptivity: The capacity of a material to absorb radiant energy. Absorbance is the ratio of the radiant energy absorbed by a body to that incident on it.

Active Residential Solar Heating System: A solar system for heating homes that utilizes forced circulation and distribution of the transfer medium. It combines the means of collecting, controlling, transporting and storing solar energy.

Air Mass: The path length of solar radiation through the earth's atmosphere considering the vertical path at sea level as unity.

Albedo: The ratio of the light reflected by a surface to the light falling on it.

Ambient: The surrounding atmosphere. Thus, ambient temperature means the temperature of the atmosphere at a particular location.

Aperture: The aperture is the opening or projected area of a solar collector through which the unconcentrated solar energy is admitted and directed to the absorber.

Automatic Damper: A device which cuts off the flow of hot or cold air to or from a room when the thermostat indicates that the room is warm enough or cool enough.

Azimuth: The angle of horizontal deviation, measured clockwise, of a bearing from a standard direction, as from north or south.

Baffle: A surface used for deflecting fluids, usually in the form of a plate or wall.

Beam Radiation: Solar radiation received from the sun without change of direction.

Black Body: A body that absorbs all incident radiation and reflects or transmits none. Additionally, a black body is a perfect radiator, i.e. it emits or radiates the maximum possible amount of radiant energy for any surface at any given temperature.

British Thermal Unit (BTU): A unit of measurement equivalent to the amount of heat energy required to raise the temperature of one pound of water one degree Fahrenheit.

Bushings: Removable cylindrical linings for an opening, such as the end of a pipe. These linings limit the size of the opening, resist abrasion, and act as a water-tight guide. Thus, when two different sizes of pipe are mated, one or more bushings might be used.

Check Valve: Used to turn off the water going through a manually controlled hot water system (or pool).

Coefficient of Linear Expansion: The change in length per unit length per degree change in temperature (centigrade).

Collection Circuit: The path followed by the collection transfer medium as it removes heat from the collector and transfers it to storage.

Collector Efficiency: The ratio of the amount of heat usefully transferred from collector into storage to the total solar radiation transmitted through the collector covers. Note: Some authorities define collector efficiency as the ratio of the amount of heat usefully collected to the total solar radiation **incident** on the collector.

Concentrating Collector: A concentrating collector is a solar collector that contains reflectors, lenses, or other optical elements to concentrate the energy falling on the aperture onto a heat exchanger or surface area smaller than the aperture.

Concentration Ratio (Concentration Factor): Ratio of radiant energy intensity at the hot spot of a focusing collector to the intensity of unconcentrated direct sunshine at the collector site.

Conduction: See Heat Transfer

Continuous Air Circulation (CAC): A mode of operation of a forced-air heating system in which the blower operates continually.

Controller, Automatic: A device used to regulate a system on the basis of its response to changes in the magnitude of some property of the system, e.g. pressure, temperature, etc.

Convection: See Heat Transfer

Cost Effective: When energy savings exceed total cost of solar system employed.

Damper: A device used to vary the volume of air passing through an air outlet, inlet, or duct.

Degree Day (DD): A unit of measurement, based on temperature difference and time, used in estimating average heating requirements for a building. For any one day, when the mean outside temperature is less than 65 degrees Fahrenheit, there exist as many DD as there are Fahrenheit degrees difference in temperature between the mean temperature and 65 degrees Fahrenheit. The base of 65 degrees Fahrenheit assumes that no heat input is required to maintain the inside temperature at 70 degrees Fahrenheit when the outside temperature is 65 degrees Fahrenheit.

Degrees Kelvin: A thermometric scale in which the unit of measurement equals one degree Centigrade, but the initial point is different. That is, the Kelvin scale is so constructed that absolute zero corresponds to $-273.16°C$. Hence, $0°C$. (the freezing point of water at sea level) is equal to $+273°K$.

Density: The ratio of the mass of a substance to its volume.

Dessicant: A substance which absorbs moisture.

Design Outside Temperature: The lowest temperature which usually occurs during the heating season at a given location. This temperature is approximately $15°F$. above the lowest temperature ever recorded by the meteorological station in the area.

Diffuse Radiation: Solar radiation received from the sun after its direction has been changed by reflection and scattering in the atmosphere.

Direct Conversion: Conversion of sunlight directly into electric power, instead of collecting sunlight as heat and using the heat to produce power. Solar cells are direct conversion devices.

Direct Gain Concept: Solar radiation is collected directly in the living space and then stored in a thermal mass.

Distribution Circuit: The path followed by the distribution transfer medium as it removes heat from storage and transfers it to the house.

Downpoint Temperature: The temperature of the distribution transfer medium as it leaves storage (below which useful heat cannot be delivered).

Drawdown: The removal of all useful heat from storage.

Duct: A passageway used for transporting air or other gas at low pressures.

Electrolysis: The use of an electric current to produce chemical changes in an electrolytic solution.

Electromagnetic Spectrum: The arrangement of electromagnetic radiations (e.g. infrared, visible, ultraviolet, etc.) on a wavelength or frequency scale.

Electroplating: The process of putting a metallic coating or plating on a base material, which is usually metal or plastic, by means of electrodeposition. That is, the metallic plating is deposited on the base by passing an electric current through a plating bath in which are dipped the objects to be plated.

Emissivity: The power of a surface to emit radiation.

Emittance: The numerical value of emissivity - expressed as the ratio of the radiant energy emitted by a surface to that emitted by a perfect radiator (black body) when the given surface and the black body are at the same temperature.

Equilibrium Temperature: The temperature of a device or fluid under steady operating conditions.

Equinox: The two times of the year when the sun crosses the equator, thereby making day and night of equal length. The spring equinox occurs about March 21 and the fall equinox about September 21.

Eutectic Material: A substance which has the property of changing from a solid to a liquid at a relatively low temperature while maintaining a constant temperature. The eutectic liquid then stores the heat energy which caused the transformation until the liquid returns to solid form and gives off heat. Eutetic salts, stored in plastic tubes, are used as thermal reservoirs to conserve and then release solar heat.

Faceted Concentrator: A focusing collector using many flat reflecting elements to concentrate sunlight at a point or along a line.

Filter: A device used to remove solid materials from a fluid.

Fin: An extended surface used to increase the heat transfer area.

Flashing: Sheet metal used in waterproofing roof valleys or ridges.

Flat-Plate Collector: A flat-plate collector is a solar collector in which the solid surface absorbing the incident solar radiation is essentially flat and employs no concentration.

Fluid: A gas, vapor or liquid.

Forced Air Heating System: A heating system in which air, circulated mechanically by either a blower or fan, is the transfer medium.

Fossil Fuel: Coal, oil, or natural gas.

Gasket: A piece of solid material, usually metal, rubber, plastic, or asbestos, placed between two pieces of pipe or between an automotive cylinder head and the cylinder block, for example, to make the metal-to-metal structure fluid-tight.

Gate Valve: Used to regulate the flow of the main supply of water.

Glazing: The transparent frontal area of flat-plate collectors.

Greenhouse Effect: The process by which sunlight is transmitted through glass and absorbed by a thermal mass; however the thermal energy reradiated by the mass, will not pass back out through the glass. (e.g. car parked in sun).

Header or Header Pipe: A section of pipe at the bottom and top of a solar collector panel; between the headers smaller pipes or channels extend in contact with the absorber. The top header carries sun-heated water away from it. A header is usually at least 1 inch in diameter and may be as much as 2-inch pipe.

Heat: The form of energy transferred from one mass to another by virtue of a temperature difference between the two.

Heat Capacity: The quantity of heat required to raise the temperature of a given mass of a substance by one degree.

Heat Exchange Flow Pattern: The relative flow arrangement of the collection and distribution circuits in storage. The typical patterns are:
1. Counterflow: collection and distribution transfer media flow in opposite directions through storage;
2. Parallel: collection and distribution transfer media flow in the same direction through storage.

Heat of Fusion: The heat released when a liquid becomes a solid, equal to the heat absorbed when a solid melts—often tabulated in units of BTU/pound.

Heat of Fusion (Phase Change) **Storage:** A heat storage medium in which the addition or removal of heat results in the medium changing state (between solid and liquid) at a constant temperature.

Heat Gain: An increase in the amount of heat in an area through direct solar radiation.

Heat Loss: A decrease in the amount of heat in an area because of loss through walls, windows, doors, etc.

Heat Pump: A refrigerating system installed where the heat discharged from the condenser is desired rather than the heat absorbed by the evaporator.

Heat Transfer: The methods by which heat may be propagated or conveyed from one place to another:
1. Conduction: Heat is transferred from one part of a body to another part of the same body or from one body to another on physical contact with it without displacement of the matter within the body. (The transfer of heat by contact with a hot body.)
2. Convection: Heat is transferred from one point to another by being carried along as internal energy with the flowing medium which can be a gas, vapor, or liquid. The two types are:
   a. Forced: results from forced circulation of a fluid, as by a fan or pump.
   b. Natural: caused by differences in density resulting from temperature changes. (The transfer of heat energy by means of the circulation.)
3. Radiation: Heat is transferred from one body to another by the passage of radiant energy between the two. The radiant energy is then converted back to internal energy when it is absorbed by the receiving body.
   a. Long Wave: radiant energy emitted from bodies at wavelengths longer than 3.0 microns.
   b. Short Wave: Radiation, Solar

Heat Transfer Coefficient: The unit-surface thermal conductance for convection or radiation which describes the rate by which heat can be transferred by either mode.

Heat Transfer Fluid: A liquid or gas that transfers heat from a solar collector to its point of use.

Heliostat: An electro-optical-mechanical device that orients a mirror so that sunlight is reflected from the mirror in a fixed specific direction, regardless of the sun's position in the sky.

Hose Coupler: A metal or plastic device for joining two pieces of hose (or one hose and a piece of rigid piping) by means of screw threads in the coupler which mate with threaded connections on the hoses or pipes.

Hot Spot: The location on a focusing collector at which the concentrated sunlight is focused and the highest temperatures are produced. If the heat is to be collected, a heat exchanger is located at the hot spot and a heat transfer fluid flowing through the heat exchanger is heated.

Hybrid Solar Energy System: A system that incorporates both passive solar design and active solar products into one building.

Hydronic Heating System: A heating system in which water is the transfer medium.

Hysteresis: The time lag exhibited by a body in reacting to changes in the forces affecting it.

Immersion-Formed Layer: A layer formed by immersion of a solid in a liquid, with the layer formation usually occurring because of electrolytic action.

Incidence Angle: The angle between the direction of the sun and the perpendicular to the surface on which sunlight is falling.

Incident: In reference to solar radiation, those light rays falling on or striking a surface.

Indirect Gain: Solar radiation is intercepted directly behind the collector by a trombe wall which serves as heat storage mass; heat energy is then transferred to the living space.

Infiltration: Air flowing inward as through a wall, crack, etc.

Infrared Radiation: Thermal radiation or light with wavelengths longer than 0.7 microns. Invisible to the naked eye, the heat radiated by objects at less than 1,000°F. is almost entirely infrared radiation.

Insolation: Sunlight, or solar radiation, including ultraviolet, visible, and infrared radiation from the sun. Total insolation includes both direct and diffuse insolation.

Insulation: A material having a relatively high resistance to heat flow and used principally to retard the flow of heat.

Isolated Gain: Solar radiation is collected and stored in a separate partition from the living space. (e.g. a greenhouse, sun porch or sunroom).

KWE: Kilowatt of electric power.

KWH: Kilowatt hour.

KWT: Kilowatt of thermal (heat) energy.

Langley: A unit of measured insolation equal to 3.69 BTU/Ft.^{2}.

Latent Heat: Heat which is necessary to produce a change of state of a material at a constant temperature.

Linear Concentrator: A solar concentrator which focuses sunlight along a line, such as the parabolic trough concentrator and the fixed-mirror concentrator.

Manifold: See Header.

MBTU: Million BTU's.

Micron: A unit of measurement of length equivalent to one millionth of a meter.

MWE: Megawatt (million watts) of electric power.

MWT: Megawatt of thermal (heat) energy.

Movable Insulation: Insulating material that can be moved and regulated according to temperature or seasonal changes to control heat loss or gain.

Optical Coatings: Very thin coatings applied to glass or other transparent materials to increase the transmission (reduce the reflection) of sunlight. Coatings are also used to reflect back to the heat exchanger infrared radiation emitted from it.

Overall Coefficient of Heat Transfer: The time rate of heat flow through a body per unit area for a unit temperature difference between the fluids on the two sides of the body under steady-state conditions.

Passive Solar Energy System: A solar energy system which operates on the principles of conduction, convection, and radiation without assistance of mechanical devices.

Phase-Change Material: A material used to store heat by melting. Heat is later released for use as the material solidifies. (See Eutectic material)

Photovoltaic Cells (Solar Cells): Semi-conducting devices that convert sunlight directly into electric power. The conversion process is called the photovoltaic effect.

Pinch Valve: A valve used to pinch off, or stop, the flow of liquid, usually water, at a desired location.

Pounds Per Square Inch Absolute (PSIA): The measurement of pressure, whether hydraulic (liquids) or pneumatic (gases), made without including the effect of atmospheric pressure, which is 14.7 PSI at sea level.

Pressure: The force exerted by a substance on a unit area of its boundary.

Primary Heating System: A system used to heat the house when the solar heating system cannot provide useful heat, usually a conventional oil, gas, or electric furnace or heat pump.

Pyranometer: An instrument for measuring sunlight intensity. It usually measures total (direct plus diffuse) insolation over a broad wavelength range.

Pyrheliometer: An instrument that measures the intensity of the direct beam radiation (direct insolation) from the sun. The diffuse component is not measured.

R Factor: Unit of thermal resistance used to compare and rate the insulating values of materials. The higher the R factor, the greater a material's insulating qualities.

Radiant Energy: The energy in the form of electromagnetic waves which is continually emitted from the surface of all bodies.

Radiation: See Heat Transfer.

Radiation, Incident: The quantity of radiation energy incident of a surface per unit time and unit area.

Radiation, Infrared (IR): Radiant energy of wavelengths longer than those corresponding to red light, i.e. longer than approximately 0.8 microns.

Radiation, Solar: Radiant energy emitted from the sun in the wavelength range between 0.3 and 3.0 microns. Of the total solar radiation reaching the earth, approximately 3 percent is in ultra-violet region, 44 percent in the visible region and 53 percent in the infrared region.

1. Diffuse sky: Solar radiation received from the sun after its direction has been changed by reflection and scattering by the atmosphere.
2. Direct (Beam): Solar radiation received from the sun without undergoing a change of direction.

Radiation, Ultraviolet (UV): Radiant energy of wavelengths from 0.1 to 0.4 microns.

Radiation, Visible: Radiation energy of

wavelengths from 0.4 to 0.76 microns which produces a sensation defined as seeing when it strikes the retina of the human eye.

Reflective Film: Transparent covering applied to the window's interior surface to block sun's heat in summer and keep interior heat from escaping in winter.

Reflective Loss: The energy which strikes a surface and is not absorbed but reflected from it.

Reflectivity: The capacity of a material to reflect radiant energy. Reflectance is the ratio of radiant energy reflected from a body to that incident on it.

Reflector: A mirror used to increase shortwave radiation input into the collector.

Resistance, Thermal: The reciprocal of thermal conductance. (See Conductance, Thermal).

Retro-fit: The installation of a solar heating system on an existing building.

Selective Black Paint: More absorbent of the infrared long wavelengths of sunlight than non-selective black paint, hence, an improved material for coating the absorber plates in solar collectors.

Selective Surface: A solar collecting surface consisting of a thin coating having a high absorbance for solar radiation and a substrata with low emittance for long wave radiation.

Sensible Heat: Heat which produces a change of temperature in a body.

Sensible Heat Storage: A heat storage medium in which the addition or removal of heat results in temperature change only (as opposed to phase change, chemical reaction, etc.). The medium is typically water or gravel (pebble bed).

Solar Altitude: the angle of the sun above the horizon.

Solar Azimuth: The horizontal angle between the sun and due south.

Solar (Photovoltaic) Cell: A device made from semiconductor materials which absorbs solar radiation and converts it into electrical energy.

Solar Collector: A device used to collect solar radiation and convert it to heat. There are two broad categories:
1. Flat Plate: Stationary and does not concentrate the solar radiation, i.e., the absorbing area is the same size as the area intercepting the incoming radiation.
2. Concentrating (Focusing): Concentrates the solar radiation incident on the total area of the reflector onto an absorbing surface of smaller area, thereby increasing the energy flux.

Solar Concentrator: Device using lenses or reflecting surfaces to concentrate sunlight.

Solar Constant: The amount of solar radiation incident on a unit area of surface located normal to the sun's rays outside the earth's atmosphere at the earth's mean distance from the sun.

Solar Declination: The angle of the sun north or south of the equatorial plane. It is plus if north of the plane and minus if south.

Solar Furnace: A unitized, self-contained, solar heating system.

Solar Glazing: A transparent or translucent covering (glass or plastic) used over windows, skylights, greenhouses and solar collectors for admitting light and preventing heat loss from reradiation and convection.

Solar Noon: The time of day when the sun is due south, i.e. when the solar azimuth is zero and the solar altitude a maximum.

Solar Window: Windows designed primarily to admit solar energy.

Solstice: The two times of the year when the sun is farthest north or south of the equator. In the northern hemisphere, the summer solstice occurs about June 21 and the winter solstice about December 21. (Summer Solstice—Longest Day; Winter Solstice—Shortest Day).

Specific Heat: The amount of heat that has to be added to or taken from a unit of weight of a material to produce a change of one degree in its temperature.

Spectral Pyranometer: An instrument for measuring total insolation over a restricted wavelength range.

Stagnation: A condition in which heat is not added to or removed from storage mechanically, but only as a result of natural heat transfer from the storage container.

Stratification: The existence of persistent temperature gradients in storage media.

Sun Factor: The average measured amount of solar radiation divided by the total possible solar radiation for a given month and location.

Temperature: Measure of heat intensity or the ability of a body to transmit heat to a cooler body.

Thermal Equilibrium: The state of a system at which there are no variations in temperature from one point to another in the system.

Thermal Lag: The amount of heat necessary to re-achieve downpoint temperature after collection is resumed following a period of stagnation.

Thermal Mass: Water drums, adobe walls or concrete floors located in a solar system for heat storage.

Thermistor: A temperature-measuring device that employs a resistor with a high negative temperature coefficient of resistance. As the temperature increases, the resistance goes down and vice versa.

Thermochemical Hydrogren Production: Use of heat with a series of chemical reactions to produce hydrogen from water.

Thermocirculation: Circulation of fluid resulting when warm fluid (liquid or gas) rises and is displaced by denser, cooler fluid in the same convective system.

Thermocouple: A temperature-measuring device that utilizes the principle that an electromotive force is generated whenever two junctions of two dissimilar metals in an electrical circuit are at different temperatures.

Thermostat: An instrument that controls temperature by responding to changes in temperature.

Ton of Cooling: 12,000 BTU/Hour. The term is derived from the amount of heat energy required to convert a ton of water into ice at 32°F. during a 24-hour period.

Total Energy System: System for providing all energy requirements, including heat, air conditioning, and electric power.

Transfer Medium: The substance(s) that carries heat from the collector to storage and from storage to the house. The medium is typically a fluid such as air, water or a water-ethylene glycol solution.

Transmissivity: The capacity of a material to transmit radiant energy. Transmittance is the ratio of the radiant energy transmitted through a body to that incident on it.

Trombe Wall: A water or masonry thermal storage wall placed directly behind the solar collector for heat storage.

Turbidity: Atmospheric haze.

Two-outlet Heater: A water heater having one outlet going typically to the piping for a domestic hot water system and another outlet, through a pipe to a large storage tank.

U Value: Heat loss through building surface, measured in BTU's per square foot for each degree difference between outside and inside temperatures.

Zenith: the point in the heavens directly over the head of the spectator; greatest height.

Appendix D:

JORDAN COLLEGE AND ENERGY

Our nation - along with the rest of the world - today faces a frightening energy crisis. So it comes as no surprise that the United States Department of Energy would spend millions of dollars in energy research. What may come as a surprise, however, is that this highly selective organization has consistently funded a small midwest Christian College with two grants of nearly $100,000 each, for solar energy, has given the college thousands of dollars worth of flat plate and concentrating collectors, and advertised its educational energy programs and energy installations all over the world. Private foundations contracted the additional thousands of dollars to build the twelve operating systems.

But one visit to the Cedar Campus of Jordan College will quickly reveal why it is being touted as a leader in the solar energy field. The Campus has twelve alternate energy installations which help to heat campus buildings and provide electricity. The College buildings, constructed of barnwood salvaged by students and staff, give the campus an earthy, rough-hewn appearance; the solar installations signify a contrasting look to the future. The young college dramatically illustrates its philosophy of "do-it-yourself" both in simple patterns of living and sophistication of scientific approach.

This small liberal arts college gained national recognition for its pioneering efforts in developing solar systems and alternate energy academic programs early in the time of our nations energy problems. More specifically, the alternate energy academic program which the College offers is a reflection of the philosophy upon which the college was created. As the school maintains the Jeffersonian principles of wedding the arts and sciences to practical application, so the technologies mirror the belief that alternate energy classes should be taught with "hands-on" experience. Thus the physical tools which textbook and theory only discuss are actually employed in the learning process.

Jordan College was founded in 1967 originally located only in Cedar Springs, Michigan, twenty miles north of Grand Rapids, the College began as a church related school. Renamed Jordan College six years later, the College became independent in 1977.

The physical plant began with one building -- a dining commons, library, and classroom complex erected in 1968. There are now seven buildings including three dormitories, a student center, a chapel and classroom building on the Cedar Campus. Most of the buildings are constructed of recycled barnwood donated by benefactors; even more impressive is the fact that students, faculty, and administrators have helped to erect the buildings at a cost cutting measure and a way to further develop the notions of conservation and self-sufficiency -- qualities which the College inspires in its courses.

Jordan College has four other campuses and offers programs through the bachelor's level. The Cedar Campus offers a four-year bachelors degree in the arts and sciences. This Campus is the only campus with resident halls and requires a Christian commitment from the students who attend.

The Jordan West Campus is the external studies program, a campus without walls and also serves five North Kent County school districts who coordinate classroom space for Jordan courses. The Newaygo county campus serves the five school districts in that county. A student may earn an Associate of Arts or an Associate of Science degree in any of almost a dozen disciplines at these locations.

The Flint Campus of Jordan College is a program developed for minority students. These students having needs different than other college freshmen are able to work with instructors sensitive to their learning requirements. The program is primarily focused on two year associate degrees with one four year program.

The Jordan Energy Institute in Grand Rapids features a concentration of five alternate energy classes taught in a one semester program. Added to that is the Associate of Applied Science degree in Alternate Energy and the Bachelor of Science in Alternate Energy degrees. Transfer programs for engineering and environmental students are articulated with several schools of engineering.

The College has been developing alternate energy education for several years, emphasizing the practical basis of research and development, with a "nuts and bolts" approach. Thousands of students have learned one or more aspects of alternate energy technolgoy in classes and seminars held throughout the Mid-west and Southern United States. Jordan College is itself the incarnation of what it teaches and values. It has been estimated that the installation of solar collectors on the Cedar Campus has had the potential of cutting fuel costs by as much as thirty percent. Through the early 80's it is projected that fuel costs will be reduced another thirty percent.

There are twelve alternate energy devices on the main campus which account for this savings. They include:

1) solar air systems
2) hot water drain-down systems for space and domestic water heating
3) glycol and silicone concentrating and flat plate systems
4) passive solar greenhouse
5) 4 KW wind energy conversion system
6) photovoltaic solar electrical system
7) wood fired hot water boiler
8) passive solar air installations

Most of the low technology solar installations on the Cedar Campus were made and installed by students and faculty. A team of professional consultants and environmental engineers from Grand Rapids and the Energy Institute supervised the construction of the higher technology systems. It should be noted that Jordan's entry into the solar era was not pre-planned. Rather its involvement in solar devices and alternate energy evolved from the College's early philosophy -- a philosophy which included two major aspects. First and foremost, it emphasized a "do-it-yourself", resourceful lifestyle in an age of great waste and overdependence upon machines, instant gratifications, and time-bought solutions. Secondly, the school in its hope to provide low cost education for students from lower socio-economic groups embraces a sense of its own responsibility in achieving this goal. Therefore, it began to seek out involvement or projects which allowed this philosophy to come to fruition.

Solar and alternate energy seemed a natural and appropriate area to pursue because it matched both dimensions of Jordan's main philosophical objectives: first, it too, emphasized conservation and allowed for a "do-it-yourself" approach; and secondly, it offered ways in which students could work on projects which not only helped to finance their education, but perhaps more importantly, served as "learning-in-process" activities for which students could gain academic credit when accompanied by in-class textbook study of the technology involved.

The first solar unit was installed in the Presidential wing of the main classroom building in 1975. Students and faculty both cut 35,000 beer cans in half, riveted them to corrugated aluminum for heat absorbers, painted them black and erected an unpretentious solar unit for classroom heating. They also assisted with the construction of thermal pane windows and the insulation of the buildings.

This project soon caught the attention of the local media. Both newspapers and television crews from all over Western Michigan began to feature the Jordan story. Public interest grew and grew, and this widespread popular interest spawned the initiation of energy classes. Other interested persons, however, indicated it would be more convenient for them to meet on weekends. Responding to this concern, Jordan College arranged no-credit mini-courses; day-long classes which packed the original course in a nutshell without skimping on essential instruction.

As demand for solar components arose from "energy" students who wished premade solar units which only needed to be assembled, Dr. DeWayne Coxon and his technicians developed unique devices which became eligible for patent application. These were a solar air absorber rod to be placed in flat plate collectors, an air handling device which has four operational modes to direct heated air from a collector, and a liquid handling apparatus which can be adapted for pool heating, domestic hot water heating and/or the heating of space.

In just a few short years, Jordan College has indeed emerged as a leader and as a great source of inspiration to individuals and institutions alike. Mutual respect and compassion for one's fellowman are its key resources. Its administrators count the intense, trying, but triumphant experiences of Jordan's brief history as tempering, yet solidifying factors. If such trials can be interpreted as enriching by its own leaders, then Jordan College need be considered truly blessed by outsiders as well. The College has already enjoyed the enthusiastic and firm support of top levels of our national government. Since its Cedar Campus has twelve operating installations, it has become a reigonal center for energy instruction.

With all these good omens -- a dedicated faculty and administration, content and grateful students, innovative programs, a physical plant which incarnates the schools' philosophy of conservation and living -- the future of Jordan College seems to insure continued success and growth. Its significance and promise, however, is perhaps best implied by these verses from an unknown author: "Do not follow where the path may lead. Go instead, where there is no path and leave a trail."

Appendix E:

Alternate Energy Installations on the Jordan College Campus

**Ms. Lexie Coxon,
Co-President**
Jordan College
Cedar Springs, Michigan, U.S.A.

Air Solar System—
Presidential Wing
(space heating)

The first building on the Jordan College campus to become solarized was the "Presidential Wing". Construction began in late 1974, and the solar system was activated in March 1976.

The "Presidential Wing," a classroom facility, has 5,000 square feet of floor space. Hydronic hot water baseboard heat supplements the solar air system.

Both the side walls and the ceiling are faced with one inch of styrofoam. In addition, the roof is of built-up design, and the side walls have R-12 fiberglass insulation. The windows are of thermopane construction and exterior doors are of insulated steel construction.

The solar installation is integral to the design of the building. The roof-mounted collector runs the length of the building, faces South at a 60 degree angle, and is divided into three sections. The collector array is 10 feet in height by 100 feet in length and is constructed of wood. Insulation is provided by high temperature styrofoam with an aluminum foil facing. The collector box is eight inches deep.

The 1000 square foot collector array is divided into three separate units, each of which heats two classrooms. Each of the three units had dissimilar metals and coverings in order to test efficiencies of different components when the collectors were first installed.

The first collector section still has a double glazed thermopane covering of 3/16 inch glass. The absorber is made of corrugated aluminum with beverage can halves riveted to it. The corrugated sheets with the attached can halves are painted with high temperature resistant flat black. There are 2¾ inches of air space between the top of the halved cans and the collector cover.

The second and third collector sections used a fiberglass cover. Both absorbers had corrugated metal with beverage cans riveted to it. The difference in the absorber was that one used aluminum components, the other steel.

Tests taking during the first three years of operation resulted in the following observation:
1. The aluminum absorber under glass activated first in the morning and activated first on cloudy days.
2. The aluminum absorber under fiberglass activated second.
3. The steel absorber under fiberglass tended to activate last and had the least efficiency.

All collectors operate at relatively the same output temperature when the insolation is 250 BTU/hr. ft.^{2}. Normal output temperatures are approximately 135 degrees at 400 CFM minimum. The maximum temperature measured was 235 degrees. More sophisticated testing will be done on all systems in the future.

The absorbers and glazing for sections number two and three were changed from the serpentine configuration to a manifold design. To accomplish this, the beverage can absorbers were removed and the Jordan hemispheric solar air rods were inserted. They were then covered with thermopane glass and the units sealed. The first heating season for the new design is 1979-1980.

Each collector section has its own heat storage area. The storage chamber dimensions are three feet wide by nine feet high by twenty-four feet long. Stones are used as the storage medium and are two to three inches in diameter. The storage chambers were oversized so that more rock could be added if more storage than originally projected would be needed. The stones are only five and one half feet high in the chambers and the total weight per chamber is 40,000 pounds.

To maximize the storage reserve capacity, the chambers were insulated first with foam glass, secondly with fiberglass, and finally with styrofoam.

Underneath the stones there are three inches of fiberglass insulation. The floors were cured well in the storage chambers before the stones were placed to minimize cracking and settling. The capacity of storage for each unit is between two and three days with the ambient temperature at 20 degrees evening and full sunlight during the day. During fall and spring weather, storage capacity time is lengthened. Generally, from the dates of December 15 to March 15 there is very little storage that takes place. Most of the solar heat is used directly from the collector into the building between those dates.

Each of the three units has an output of approximately 50,000 BTU's per hour. When sun is not available, the supplemental hydronic baseboard heating system automatically activates to heat the building. The energy saving is approximately 35-40% of annual heat load requirements.

Cost calculations are difficult. The total unit is indigenous to the building and reclaimed barn lumber was used for much of the construction. It is generally felt that the solar application added $7,000 to the total cost of the building. This includes the cost of building the collectors and storage and installing the electrical and mechanical equipment.

With the rising cost in energy, Jordan College believes it was a wise idea to build the solar system on the "Presidential Wing". All new buildings on the Cedar Campus are presently being planned with alternate energy systems as an integral part of their energy efficiency.

Hot Water Drain-Down System— (DOE)—Retrofit—Madison Hall
(space heating and domestic hot water)

The second major solar installation on the Cedar Campus was a retrofit on Madison Hall. Funding for this $138,000 project was supplied by the Department of Energy (DOE) and Jordan College. The College was fortunate to be chosen in the second round of solicitations through the DOE Program Opportunity Notice, which is their solar demonstration program.

Jordan College had requested funds from DOE for its first solar project but was refused because the system was not duplicable. It was after being informed by DOE that commercially available components must be used that the engineering firm of Fairbrother & Gunther, Inc. of Grand Rapids was asked to join the Jordan alternate energy technical team for a joint venture in retrofitting Madison Hall. Madison Hall, a dormitory and lounge, has 7,250 square feet of space, and houses 60 students. There is also a dean's apartment and a student lounge.

Built in 1970, energy conservation techniques were utilized throughout, including quality insulation and thermopane Anderson windows. The retrofitted solar system is connected to the 300,000 BTU gas fired hot water boiler. Utilizing baseboard heating, the system has a constant 180 degree heating zone with a pump and burner cycled by demand, with zone controlled thermostats. The domestic hot water heater is a 199,000 BTU gas fired unit with 100 gallons storage.

A drain-down system or a high temperature, non-freezing fluids system were the options. Because the building demanded 180 degree heated water for the baseboard radiation on the coldest winter days, it was necessary to use a system that did not have heat exchangers and would move the water directly from the solar collectors into the hydronic system for maximum heat gain. The system uses no anti-freeze to protect it from freezing.

The collectors for the Madison Hall installation are Pittsburgh Plate Glass model B, glaze, copper absorber plate with flat black coating and copper internal pipe. There are 104 collectors for a total net square footage of 2,080. The collectors are tilted at a 60 degree angle facing the south and are stationary.

Water is pumped from the mechanical room into the supply lines, up through the collectors to the headers, and back into the mechanical room. The pump is a high velocity Bell and Gosset, and the water is taken out of the bottom of the "day tank".

The mechanical room houses three types of storage units:
1. The day tank is a three hundred gallon unit that holds the water which is cycled every eight minutes through the collectors and back to the mechanical room.
2. The main storage tank is a 5,000 gallon unit that receives the surplus solar heated water after the day tank has reached a temperature of 180 degrees.
3. The solar domestic hot water heaters are two 80 gallon tanks that receive water before

it goes into the gas fired domestic hot water heater.

In addition to these three water storage units, two drain down (ballast) tanks are mounted directly over the day tank. These two 85 gallon units are the expansion tanks that receive the drain down water when the system turns off and contain the nitrogen that is used to purge the collectors when the water is pumped out. These maintain the liquid and nitrogen combination that make the system an automatic drain down type.

The 104 collectors have a total BTU output of approximately 500,000 per hour. That output is at an insolation value of 240. Calculations indicate that 25% of the space heating load and 50% of the water heating load will be provided by this Solar System.

The drain down feature of the collectors, operating under 15 pounds pressure, is a feature that is quite uncommon in solar demonstration units. In northern climates there is a start-up lag time if liquids are left outside the building because they are super-chilled by a negative ambient temperature. Drain down systems resolve that by keeping the fluids inside the insulated containers.

Jordan Air Solar System—Retrofit —Jefferson Hall
(space heating)

Jordan College was teaching a do-it-yourself solar air system early in the history of the Energy Division. After several hundred students had taken both the Solar I academic class and the solar mini-course "Here Comes the Sun," there was a need expressed to have a residential demonstration site with the Jordan Air System on it. Additionally, there was a need for a slide demonstration to be used in the courses which would help students understand construction of the total Jordan System.

Jefferson Hall is a 70 year old house on the Northeast corner of the campus. It is of wood frame construction and is presently used as a women's residence building. It has approximately 1,200 square feet of living space and is a two story structure.

The building is heated by a converted coal furnace that has been adapted to natural gas. The coal bin had been changed to a fruit cellar.

Two collectors were mounted on the south facing roof of the building.
1. One collector is what is described as a "Jordan Collector". It is an eight by twenty-four foot wooden box that was fabricated from materials purchased from the local lumber dealer. The box is constructed of pine, but many students have since used redwood or wolmanized (pressure treated) wood. Placement of the structural supports for air flow patterns was completed as noted in the Jordan manual "Here Comes the Sun". Jordan air absorber rods, of anodized aluminum extruded material were installed in the collector box. A fiberglass cover with a transmission factor of 82% was used. The collector is 196 square feet and has an insolation factor of 240. The system will generate 32,000 BTU's per hour. The collector tilt is 60 degrees, it faces due south, and is in a stationary position.
2. The other collector is a six by twenty-four foot wooden box which was constructed for the first Jordan Solar classes. It used the aluminum beverage can air design similar to that used on the "Presidential Wing". The cans were cut in half, riveted to the corrugated aluminum backing and painted black. There are fiberglass covers on this unit also. The collector faces due south at a 45 degree angle. Its total output is 23,000 BTU's per hour.

The purpose of raising one collector to 60 degrees and elevating the other to 45 degrees was to contrast the BTU output over a heating season. Comparative testing had to be discontinued during the winter months due to the snow buildup that took place on the second collector tilted at 45 degrees. The 60 degree angled collector had no problem with snow buildup. The 45 degree tilted collector had continual difficulty with snow sticking to it which permitted only occasional data collection. Jordan technicians felt that the solution to the problem was to raise the collector to the 60 degree position, which is the normal position for collectors at this latitude.

The coal bin in the basement of Jefferson Hall was converted to solar storage. It was well insulated and filled with approximately 202 cubic feet or about 15,000 pounds of rock. The total storage output is 55,000 BTU's per hour. The cost of the system with minimum controls and no air handler was approximately $1200. The calculation for energy saving on the building is approximately 30% annual average.

Hot Water Drain-Down System— Prototype—Chapel
(domestic hot water)

While research was being done in water handling apparatus for students, a demonstration prototype of the Jordan water handler was installed on the College chapel for development purposes. The solar water handler, a fully automatic unit that can be used by the solar do-it-yourselfer, can be used for domestic hot water pre-heat, space heating, and pool heating.

The collector for this system is constructed of Mueller Brass D-tubes soldered to copper sheets. The surface was then painted black and glazed with thermopane glass. The collector is mounted at a 50 degree angle for better summer sunlight exposure. It has 196 square feet of available surface and will generate approximately 47,000 BTU/hr. at an insolation factor of 240 BTU/ft^2.

The storage tank has a capacity of 68 gallons. The pump is automatically programmed through its modes by the hot water demands of the building and the availability of the sun. Four zone valves are programmed to route the water from day tank to collector; day tank-to collector-to storage; or day tank-to-storage-to day tank. They operate in coordination with one pump. The zone valves have fail-open safety features so that during a power loss all water will be able to drain back into the day tank from the collector. Such features are necessary to protect against freezing.

Hot Water Drain-Down Solar System—Ford Administration Building
(domestic hot water)

The Ford Administration building domestic hot water needs for the kitchen and restrooms are preheated by a drain-down PPG Solar System. That system consists in three panels mounted on the roof facing South placed at a fifty degree angle to the sun.

The system has approximately fifty square feet of exposed collector; the absorber is of copper plate with brazed copper water lines on six inch centers; and glazing is double pane with two inch fiberglass insulation.

Storage is provided by a fifty gallon tank, powered with a Grunfos pump. The valves are solenoid with fail open and fail closed features. The pump is activated when the sensor in the collector registers a higher temperature than the sensor in the tank. Water drains externally from the system when collector temperature drops below forty degrees. The pump ceases to operate when storage temperatures rise above two hundred degrees. No anti-freeze is used in the system.

Passive Solar Greenhouse—Lincoln Lounge

Forty-eight inch deep footings support external walls which bear the weight of triple glazed 3/16" thick tempered glass for the greenhouse. Adjacent to the student lounge, this 14 x 26 foot structure has survived the Michigan winter without freezing.

An automated solar blanket, with reflective coating for summer use, will serve to help retain heat collected and stored in eutectic salts during winter evenings and on overcast days. By triple glazing, insulating the foundation with Dow sheathing, and being cognizant of other passive design features, the greenhouse requires only a small amount of supplemental heat during the most severe portion of the winter to sustain vigorous plant growth.

Power for lighting is supplied by the WECS as direct current (DC) with no inverter.

Wind Energy Conversion System (WECS)—4.5 KW

The Earl-Beth Foundation of Detroit has participated in innovation projects for many years. Its president, Danforth Holley, has often contributed "seed money" for pilot projects such as the Jordan WECS. Upon funding from the foundation, a 4.5 KW machine was purchased from Dakota Sun and Wind, Inc., and an oil service derrick was purchased from the Central Michigan oil fields.

The derrick is 92 feet tall, has four pedestals and rests on concrete pilings six feet deep and two feet in diameter. The custom built stub tower is bolted to the main tower by large steel bolts through heavy steel angle iron. The generator is one of the first new units built from the original Jacobs design. Components have been enlarged in order to produce a maximum 4 KW at wind speeds of 24 miles per hour. The unit has three six foot blades and is an up-wind machine. The tail assembly points the machine into the wind and maintains such a position until gale forces allow for the feathering of the blades through spring loaded action. Blades return to their former position as the wind subsides.

Storage for the wind generated electricity is provided by fifty-five two-volt batteries which are located at the base of the tower. No inverter is used, and the direct current is used to power DC motors and lights for the passive solar greenhouse and Lincoln Lounge.

Thermosyphon Solar System—Cedar Chapel

Two Revere Collectors consisting of copper plate on copper lines comprise 50 square feet of absorber for this system. During the summer the thirty gallon storage tank is drained of its antifreeze solution and stored. Water is then added for the summer.

The storage tank is a thirty gallon container placed 12 inches above the collectors and by molecular expansion the liquid rises through the collector into the storage tank. The solar gain is then transferred to domestic use through a heat exchanger in the storage tank of ¾ inch copper coil.

The Futuristic Look

Jordan College is committed to the exploration and utilization of alternate energy. A wood burning boiler is presently being installed to assist in heating the buildings when the sun is not available. As the Cedar Campus receives the most attention due to its high concentration of solar and wind systems, it has been possible to attract more government participation by a donation of 40 Northrup concentrating collectors for immediate installation.

A current proposal to the Department of Energy to install a concentrating solar system on the Ford Administration Building that will both heat and cool the facility is presently under review in Washington. That $260,000 system would be the first of its kind in mid-America.

Completing the total energy profile of the Cedar Campus is the Environmental Resource Center, a utility-free structure costing about 1.5 million dollars. Not only will it be the model energy building for today, but it will rest atop a 200,000 gallon solar sink that will store, on an annual basis, the BTU's from the summer sun to be used in the winter heating season. All campus solar systems will feed into a loop carrying the water to the solar sink.

The Environmental Resource Center (ERC), submitted under the Department of Energy's PON 1979 Cycle Four will utilize both passive and active solar. The 80 foot high diamond shaped solar tower connecting the two story library integrates with a circular earth bermed administrative complex that will be lighted with a solar heliostat. Photovoltaic collectors mounted on the south facing roof will assist the low temperature solar powered turbine in providing electricity for the building and the campus. A twenty-foot Darius wind machine on the peak of the solar tower will generate electricity for non-sun days.

The building of the Environmental Resource Center in 1981 will enhance the energy independence for the Cedar Campus, and make Jordan College an archetype for the battle of small colleges against rising energy costs.

Appendix F:
Solar Tax Credits

For CERTIFIED solar, wind, or water energy conversion systems, the credit will apply to systems installed between January 1, 1979, and December 31, 1983. The percentage of the costs which can be credited against income taxes declines yearly after 1980. The credit for sing-family residences are:

Year of Installation	Percentage of First $2,000 of Cost	Percentage of Next $8,000 of Cost
1979	25	15
1980	25	15
1981	20	10
1982	15	5
1983	10	5

For residential buildings other than single-family declines yearly after 1980. The credit for single family residences are:

Federal Income Tax Credits (tax credit means a direct reduction in tax liability). On the first $2,000 expended by the homeowner, the Federal government will grant 30% or $600. On the next amount, from $2,000 on up to $10,000, the Federal government will grant a 20% tax credit or $1,600, for a total Federal Income Tax credit of $2,200 on a system costing up to $10,000. At the present time this applies only to one's domicile and the regulations promulgated by the Internal Revenue Service all but rule out passive systems. Their regulations with regard to passive are very restrictive, i.e., if you have a trombe wall and it is load bearing, it does not qualify for a tax credit. If the glazing on the South wall also serves as a window, it too is disqualified for tax credit purposes.

State Solar Tax Credits

ALASKA—A tax credit amount to 10 percent ($200 maximum) of the cost of installing a solar system in the taxpayer's personal residence is available. The credit applies to expenditures incurred between January 1, 1977 and December 31, 1982.

ARIZONA—The cost of solar devices on all types of buildings can be amortized over 36 months in computing net income for State income tax purposes. Income tax credit amounting to 35 percent ($1,000 maximum) of the cost of installing a solar device in a taxpayers residence is also available. The credit percentage declines five percent a year until the law expires in 1984.

ARKANSAS—Individual taxpayers are allowed to deduct the entire cost of solar heating/cooling equipment from gross income for the year of installation.

CALIFORNIA—An income tax credit amounting to 55 percent ($3,000 maximum) of the cost of installing a solar system in a home is available. The installed systems must meet criteria established by the Energy Resource Conservation and Development Commission. For any building other than a house, where the cost of the system exceeds $6,000, the credit provided is $3,000, or 25 percent of the system cost, whichever is greater. The law expires on January 1, 1981.

COLORADO—A personal and corporate income tax deduction is available equal to the cost of installing a solar energy device in a building.

CONNECTICUT—Sales and use tax exemptions are provided for solar collectors. Exemptions expire October 1, 1982.

GEORGIA—Sales and use tax refunds are provided for the purchase of solar equipment.

HAWAII—An income tax credit amounting to 10 percent of the cost of a solar system is available. The system must be placed in service between December 31, 1974 and December 31, 1981.

IDAHO—The entire cost of installing a solar system in a residence can be deducted over a four year period. Deductions cannot exceed $5,000 in any taxable year.

KANSAS—Individuals and businesses are allowed an income tax credit equal to 25 percent of the cost of a solar system (up to $1,000 for individuals and $3,000 for businesses). If individual credit exceeds tax liability, excess credit can be carried forward for 4 years. The system must be acquired prior to July 1, 1983. In addition, the taxpayer can amortize and deduct the cost of a solar system installed in a business over 60 months.

MAINE—Provides a refund of sales or use tax paid on solar equipment certified by the Office of Energy Resources. The provision expires on January 1, 1983.

MASSACHUSETTS—A corporation may deduct the cost of a solar heating system for taxable income. There is also an exemption of retail sales tax from solar systems used in an individual's principal residence.

MONTANA—An income tax credit is provided for installation of a solar system in taxpayer's residence prior to December 31, 1982. The credit amounts to 10 percent of the first $1,000 spent and 5 percent of the next $3,000.

A personal or corporate tax deduction is also available on the capital investment for a solar system. Maximum deduction for a residence is $1,800 and $3,000 for a nonresidential building.

NEW JERSEY—Solar systems are exempt from sales and use taxes. To qualify for exemption, the system must meet the standards set by the Division of Energy Planning and Conservation, State Department of Energy.

NEW MEXICO—An individual income tax credit for 25 percent ($1,000 maximum) of the cost of a solar heating/cooling system installed in a residence or used to heat a swimming pool is available. If the credit exceeds tax liability, a refund is paid. A credit cannot be claimed if the taxpayer claimed a similar credit, deduction, or exemption on his Federal income tax return. The solar system must meet performance criteria prescribed pursuant to the Federal Solar Heating and Cooling Demonstration Act of 1974.

NORTH CAROLINA—A personal and corporate income tax credit is available equal to 25 percent ($1,000 limit) of the cost of a solar heating/cooling system installed in a building. The system must meet performance criteria prescribed pursuant to the Federal Solar Heating and Cooling Demonstration Act of 1974.

NORTH DAKOTA—An income tax credit, equal to 5 percent a year for 2 years, is provided for installation of a solar system.

OKLAHOMA—An income tax credit for 25 percent ($2,000 limit) of the cost of a solar system installed in a private residence is available. This law expires on December 31, 1987.

OREGON—A personal income tax credit for 25 percent ($1,000 limit) of the cost of installing a solar system in a home between January 1, 1978 and January 1, 1985 is available. The system must provide at least 10 percent of the home's energy requirement and must meet performance criteria adopted by the State Department of Energy.

TEXAS—An exemption from sales tax is provided on receipts from sale, lease and rental of solar devices. A corporation may also deduct the amortized (60 months) cost of solar system by the corporation from taxable capital.

VERMONT—Personal and business income tax credits are available for installation of solar systems prior to July 1, 1983. Credit available is equal to the lesser of 25 percent of the cost of the system installed on real property, or $1,000 ($3,000 for businesses).

WISCONSIN—Businesses may either deduct (in the year paid or depreciate or amortize over 5 years) the cost of installing a solar system. Also, an income tax credit is available to individuals for installation of a solar system. Amount of the credit depends on the installation year and the date the structure, to which the solar system is attached, appeared on the tax toll. If the credit exceeds tax liability, a refund is paid. The law applies to expenses incurred between April 20, 1977 and December 31, 1984.